ADVANCES IN
THE ECONOMICS OF
ENERGY AND RESOURCES

Volume 7 • 1992

ENERGY, GROWTH, AND
THE ENVIRONMENT

ADVANCES IN THE ECONOMICS OF ENERGY AND RESOURCES

A Research Annual

ENERGY, GROWTH, AND THE ENVIRONMENT

Editor: JOHN R. MORONEY
Department of Economics
Texas A&M University

VOLUME 7 • 1992

 JAI PRESS INC.

Greenwich, Connecticut   London, England

LAMAR UNIVERSITY LIBRARY

Copyright © 1992 by JAI PRESS INC.
55 Old Post Road, No. 2
Greenwich, Connecticut 06836

JAI PRESS LTD.
118 Pentonville Road
London N1 9JN
England

All rights reserved. No part of this publication may be reproduced, stored on a retrieval system, or transmitted in any form or by any means, electronic, mechanical, photocopying, filming, recording, or otherwise, without prior permission in writing from the publisher.

ISBN: 0–89232–947–5

Manufactured in the United States of America

CONTENTS

LIST OF CONTRIBUTORS	vii
INTRODUCTION *John R. Moroney*	1
FINDING AND DEVELOPING COSTS IN THE UNITED STATES, 1945–1986 *M.A. Adelman*	11
THE PETROLEUM PROBLEM: THE INCREASING NEED TO DEVELOP ALTERNATIVE TRANSPORTATION FUELS *D.J. Santini*	59
ENERGY AND MACROECONOMY: CAPITAL SPENDING AFTER AN ENERGY COST SHOCK *D.J. Santini*	101
REDUCING U.S. CARBON DIOXIDE EMISSIONS: THE COST OF DIFFERENT GOALS *Dale Jorgenson and Peter J. Wilcoxen*	125
DYNAMIC MODELS OF INPUT DEMANDS: A COMPARISON UNDER DIFFERENT FORMULATIONS OF ADJUSTMENT COSTS *G.C. Watkins and Ernest R. Berndt*	159
ENERGY, CAPITAL, AND GROWTH *John R. Moroney*	189
ENERGY CONSUMPTION, CAPITAL, AND REAL OUTPUT: A COMPARISON OF MARKET AND PLANNED ECONOMIES *John R. Moroney*	205

LIST OF CONTRIBUTORS

M.A. Adelman — Department of Economics
Massachusetts Institute of Technology

E.R. Berndt — Sloan School of Management
Massachusetts Institute of Technology

Dale W. Jorgenson — John F. Kennedy School of Government
Harvard University

John R. Moroney — Department of Economics
Texas A&M University

D.J. Santini — Center for Transportation Research
Argonne National Laboratory

G.C. Watkins — Data Metrics Limited
Calgary, Alberta, Canada

Peter Wilcoxen — John F. Kennedy School of Government
Harvard University

This volume is dedicated to
the memory of
Joseph J. Spengler,
an inspiration to
all who knew him

INTRODUCTION

John R. Moroney

This volume features two broad themes: energy and economic growth, and energy and the environment. The papers by Moroney and one by Santini, as well as the Jorgenson–Wilcoxen paper, deal with energy and economic growth. A second paper by Santini centers on the feasibility of adopting methanol as a widely-used fuel in automobiles and other segments of transportation. Watkins and Berndt focus on the costs of adjusting a quasi-fixed input, capital, to long-run equilibrium in dynamic models of multi-input demand, which include energy, labor, and materials.

Adelman's chapter chronicles the real costs of discovering and developing oil reserves in the United States, 1945–1986. It is no secret that we face a long-term problem of physical depletion of domestic reserves. As of 1985, Adelman reports that 60 percent of the largest fields discovered in the United States had already been found before 1930. And 186 of these largest fields contained about 80 percent of all the oil that had been discovered in large fields before 1945. Using 1959–1960 as a benchmark, Adelman shows that unit costs of new reserves added were roughly $3–4 per barrel (U.S. $1983). But by 1985–1986 these costs had risen to roughly $4–6 per barrel. Annual observations on the cost of new reserves are quite erratic, but a sharp upward trend is obvious from 1972 to 1985–1986 for reasons Adelman spells out clearly. The same erratic pattern and trend hold for costs per million cubic feet (Mcf) of new non-associated gas reserves. But recorded gas reserves are highly questionable because of major year-to-year error in reported reserve additions, as Adelman emphasizes. Since development investment and exploratory investment are substitutes, development cost trends, which can be measured, are a surrogate for exploratory cost trends, which cannot. Another surrogate for real exploratory cost is the trend in physical reserve values, which also parallels development cost.

A major contribution of Adelman's paper is in showing that publicized accounts of replacement costs in 1984 of $26–27 per barrel of new oil reserves are drastically overstated, perhaps by a factor of 5 or 6. Yet depletion remains a vexing problem. Although the United States experienced a surge in exploratory and development drilling in the mid and late 1970s, comparatively modest oil and gas reserves were added. In fact, the troublesome puzzle he states, but does not resolve, is: Why, with sharply higher real prices, were reserve accretions so abysmally bad in the early to middle 1970s?

Adelman concludes on a cautiously optimistic note. He feels that development costs per barrel of oil must inexorably creep upward, but he is not alarmed by the rate at which they are likely to increase. Despite Adelman's carefully tempered view on costs, domestic production of oil has declined from about 9.5 million barrels per day in 1970–1972 to 7.3 million in 1990. By contrast, 1990 U.S. consumption of crude and refined oil was about 16.9 million barrels per day. There is still ample time, but the moment has arrived to begin a determined search for alternative fuels in this country. The time will be squeezed if major war bursts in the Middle East.

This fact leads to Danilo Santini's paper emphasizing the need to develop alternative fuels in the U.S. transportation sector. Santini stresses an important point: The transportation sector currently accounts for over 60 percent of oil consumed in the United States. The situation is only slightly better in the remainder of the OECD countries, where nearly 40 percent of total oil consumption is attributable to their transportation sectors. Santini further stresses that consumption technology in transportation sectors promises very limited short-run and even long-run price induced substitution against gasoline. The brightest hope may well rest in the development of broadly-accepted substitutes for gasoline.

Santini sees methanol as an economically and technologically viable substitute. On the supply side, Santini documents the feasibility of producing methanol from natural gas. He further emphasizes the desirability of methanol as a clean fuel for the future. Worldwide natural gas reserves are increasing at a far faster rate than worldwide oil reserves. Hence gas reserves offer a longer-term resource base from which methanol could be developed as a transportation fuel for at least several decades. On the demand side, Santini emphasizes the feasibility of using methanol in automobiles (and transportation more generally) with some engine materials modification. This view is supported as well by ranking executives of a few major oil companies and General Motors Corporation.

Santini cautions that widespread use of methanol would not be simple or cheap. Methanol has three main drawbacks. First, it is highly poisonous to primates. It can be toxic to humans in doses as small as three or four teaspoons; therefore, because it is odorless and tasteless, it would need to be clearly identified before being distributed widely. Second, methanol contains only about half the energy per gallon as gasoline. Because methanol burns somewhat more efficiently than gasoline does not nearly offset its lower energy content, and its volumes would be large. Third, methanol is more corrosive than gasoline or oil. Hence marketing would require new refueling pumps, storage tanks made of carbon steel or special fiberglass, and possibly new pipelines. Given these drawbacks, a study funded by oil companies in California estimates that an 85 percent methanol blend could feasibly be marketed for roughly $0.83 to $1.00 per gallon, which translates to a "gasoline equivalent" retail price of about $1.40 to $1.65 per gallon (in 1988 dollars).

In December 1990, General Motors has developed an engine that will accept methanol, gasoline, or any mixture of the two. Chrysler has also produced a methanol-burning engine that is capable of using the highest-octane straight gasoline. Another alternative fuel, to be marketed as Americlean, was announced in December 1990. Americlean is a liquified natural gas at customary ranges of temperature, so long as it is maintained under sufficient pressure. Apparently, it can be used in several current car engine designs. A pilot plant in Alabama is reportedly producing roughly 1 million gallons per week. This may develop as another environmentally benign alternative to gasoline.

Global warming may become a serious problem conceivably in the next 40 or 50 years. The recent Clean Air Act poses binding environmental constraints on permissible carbon and nitrogen emissions in the United States over the next 15 years. Carbon dioxide is the most important contributor to the greenhouse effect. Other important contributors are methane, chloroflourocarbons, ozone, and various oxides of nitrogen. One means of reducing greenhouse gas emissions is to tax their sources. The most important is carbon.

Jorgenson and Wilcoxen develop a 35–sector general equilibrium model of the United States to analyze the economic impact of taxing carbon fuels. Because energy is one source of labor productivity growth (along with physical capital and innovation), restricting carbon emissions constrains potential productivity growth. The degree of growth retardation, in turn, depends inversely on the substitutability between other inputs and carbon fuels.

The regulation strategy considered by Jorgenson and Wilcoxen has two parts: (1) a target path of emissions and (2) an instrument used to attain the target. Jorgenson and Wilcoxen consider three targets: (1) immediately stabilize carbon emissions from fossil fuels at 1990 basecase levels (1,576 million tons); (2) decrease carbon emissions gradually during the period 1990–2005, reaching 80 percent of the 1990 level in year 2005 and maintaining it thereafter (at 1,261 million tons); and (3) do nothing until the year 2000, then tax fuels so that in the year 2010 carbon emissions are stabilized at the projected year 2,000 level (1,675 million tons). The three fuels are coal, oil, and natural gas.

As one might expect, the carbon emissions rate per btu is lowest for gas, with oil having an emissions rate 50 percent higher than gas, and coal a rate about double that of gas. For this reason, a tax designed to achieve a given level of aggregate carbon emissions sharply raises the real cost of coal and constrains its use. In their translog model simulations, other fuels are substituted against coal. And because a carbon tax raises the cost of energy as an aggregate, there ensues general substitution of materials, labor, and capital against energy. Interestingly enough, taxing fuels so as to maintain the 1990 carbon emissions level has a negligible impact on most sectors and aggregate economic growth: By 2020, GNP falls by only one-half of one percent, relative to the base case.

If emissions are to be reduced to 80 percent of the 1990 level, macroeconomic consequences are much more serious. To achieve this goal requires a tax of $60.09 per ton of contained carbon, which is equivalent to a tax of $39.01 per ton of coal, $8.20 per barrel of oil, or $0.98 per Mcf of gas. Such a tax would devastate the coal industry. The price of coal would more than double and output would fall to only 47 percent of base-case level. By the year 2020, GNP would decline by 1.6 percent relative to the base case.

Given the Jorgenson–Wilcoxen model, these are huge costs for achieving a marginally cleaner environment. And the United States, after all, is not the principal producer of coal carbon emissions—China is. It seems to me quite doubtful that a stringent tax on fossil fuels levied solely by the United States would perceptibly diminish the greenhouse effect globally. If this be so, then the sacrifice in potential domestic growth from such taxes seems too costly. In my view (and this is quite apart from the Jorgenson–Wilcoxen analysis), to reduce the rate of global carbon emissions by, say, 20 percent or 30 percent from 1990 levels will require binding multinational restrictions. Feasible political mechanisms for enforcing such potential restrictions strain one's imagination. In any case, substantive reductions by the United States alone cannot solve the global problem.

Energy consumption, capital formation, and technological progress are the cornerstones of growth in all industrial economies, East or West. It is no easy business to account for their individual effects, but their combined importance is indisputable. The macroeconomic channels connecting energy use, capital accumulation, and assimilation of new technology are exceedingly complex. Santini's major thesis in "Energy and the Macroeconomy" is that energy price shocks are a key determinant of business cycles and long-term economic growth. Jevons, Keynes (particularly in *Economic Consequences of the Peace*), Hayek and, of course, Georgescu–Roegen have stressed the importance of energy consumption governing an economy's long-run production potential.

Santini's short-run model works as follows. Energy price shocks set in motion microeconomic incentives for firms to reduce energy use relative to other inputs, including capital. Santini focuses on cash flow constraints, caused by energy price increases, that limit a firm's *ability* to purchase new, energy-efficient capital. The cash flow constraint causes a short-run reduction in actual investment spending that induces a transitional recession.

Santini argues correctly that energy and physical capital are long-run substitutes. Yet they may appear to be complements in time series studies. His central idea is straightforward. Consider a firm that employs capital and energy only. An energy price shock reduces immediately the quasi-rents accruing to capital (Santini calls this the firm's capital payment possibilities). The abruptly higher energy price thus spurs the firm's wish to substitute new, thermodynamically efficient capital for energy but simultaneously impairs its ability to do so. Firms whose revenues cover their variable (energy) operating costs stay in business, but face financial strictures that reduce their ability to purchase new, energy efficient capital. Firms then extend for a time the life of current capital stocks. In such short-run disequilibrium, an increase in real energy price reduces the firm's and the aggregate short-run effective demand for capital and makes physical capital and energy appear to be complements when in fact they are long-run substitutes. Santini's main contribution is in showing why short-run maladjustments attributable to financial constraints can be more serious than long-run comparative static cost minimization implies.

Watkins and Berndt analyze adjustment costs using a four input (capital, labor, energy, and materials) restricted variable cost function. They analyze a "third generation" dynamic model specifically as estimation of the variable cost function is embedded in the adjustment path of the quasi-fixed factor to its new equilibrium. Physical capital is treated as a quasi-fixed input, the others as variable. Their goal is to analyze the time path and costs of adjusting capital from one equilibrium state to another. They center on whether the specification of adjustment costs is better depicted by changes in net or gross investment.

The costs of adjusting the capital stock due to net investment are straightforward. However, they emphasize that even if net investment were zero (so that all gross investment were strictly to replace retired capital), adjustment costs would nonetheless be positive; pure replacement investment is not frictionless. Internal costs of adjusting to the optimal capital stock are firm specific, but the Watkins–Berndt samples are of industries rather than firms. Their model is applied to three time series samples: Canadian iron and steel, Canadian textiles, and aggregate Canadian manufacturing, 1957–1982.

Empirically, the key differences between the two adjustment cost formulations are twofold. First, if adjustment costs depend chiefly on net investment, the lagged capital stock does not appear in the derived input demand equations. But if adjustment costs depend mainly on gross investment, the lagged capital stock does appear in these equations.

Introduction

Second, the investment function is more complex in the gross investment model because it includes replacement investment.

Application of the two models produces mixed results. For aggregate Canadian manufacturing, the gross investment model is superior in one important sense: The net investment model yields coefficient estimates that violate second-order stability conditions. However it is disappointing in that the gross investment demand equation carries no statistical significance, whereas the net investment demand equation does. Inferences concerning long-run substitution and complementarity depend on which adjustment cost model is used. In the net investment model parameter estimates indicate capital and labor to be complements, while in the gross investment formulation they appear to be substitutes. By contrast, capital and energy are long-run substitutes in the net investment model, and complementary in the gross investment model.

Application of the two models to the iron and steel industry yields notably similar results with only one real disparity. Labor and capital show up as substitutes in the net investment model but complements in the gross investment formulation. Further, the gross investment model produces a substantially larger log likelihood value for the system of equations.

When applied to the textiles industry, the gross investment model again has a notably higher log likelihood value for the system. There is one notable anomaly—estimated coefficients imply that capital and materials are long-run substitutes in the net investment model but complements in the gross investment model.

In my view, these differences spotlight a truly vexing point. Comparing capital with other inputs, the statistically inferred long-run substitution and complementarity relationships are sensitive to the assumed source of investment costs. However this should not be so: capital adjustment costs may quite properly govern the speed the firm adjusts capital with respect to other inputs. But the long-run production function and nature of input substitution or complementarity are technological facts independent of the source of adjustment costs. Our inability to sort out these facts about substitution and complementarity in such econometric work rests not in the theory or in model specification.

The Watkins–Berndt models are quite rigorous and are estimated by entirely proper methods. Instead, the truths elude us because these time series (and most other aggregative data sets) are not rich enough in information to produce robust parameter estimates using such complex econometric models. Put another way, model specification can overburden the data. If so, our models may yield contradictory (yet each statistically precise) point estimates of the true parameters. Without richer data, it may be better to use simpler models.

In "Energy, Capital, and Growth," Moroney analyzes a two-equation model of economic growth for five OECD countries (France, Germany, Sweden, Canada, and the United States) from the 1950s through 1980. Growth trends in each country are distinguished by two epochs: pre–1973 (before the first major energy price shock) and 1973–1980. Using the Summers–Heston data for real per capita income, Moroney finds that growth rates in all countries fall abruptly after the price shock. Through 1973 each country displayed nearly identical internal growth rates in physical capital and energy consumption per capita, and in real per capita income. These rates range from 4 or 5 percent in France and Germany to 3 or 4 percent in Sweden and Canada to 2 percent in the United States. Each country experienced internally balanced substitution of energy and capital for labor, which by itself (without knowledge of input prices) could indicate capital–energy complementarity.

After the 1973 energy price shock, the internally balanced substitution and real per capita income growth change abruptly. First, per capita energy consumption shifts from strong positive growth to actual decline in all countries except Canada (where growth falls from 3.3 to 1.1 percent). Second, however, real capital stocks per capita continue to grow in all five countries (and accelerate in France, Canada, and the United States). Although rates of capital utilization declined following the price shock (and are not taken into account here), capital stock accumulation per capita continues apace in each country.

Most importantly, growth rates in real per capita income fall sharply after 1973. Moroney attempts to separate the contributions of capital and energy to per capita income growth. He concludes that energy consumption contributed approximately one-third and capital accumulation approximately two-thirds to the actual growth in real per capita incomes in most countries. He also estimates a long-run distributed lag demand equation for energy. Estimated long-run price elasticities of demand are in the range –0.25 to –0.45, while income elasticities of demand are approximately one in all countries.

How do energy per worker and capital per worker stack up in East and West Europe? Since World War II, both the market-type economies of the West and the former East European members of the Soviet Bloc

Introduction 9

vastly increased both energy and capital per worker. In "Energy Consumption, Capital, and Real Output: A Comparison of Market and Planned Economies," Moroney finds that for the years 1978–1980 East Europe had a substantially lower range of real GDP per worker than West Europe per worker (using real GDP estimates from the International Comparison Project). Yet East Europe used on average more energy, but less capital, per worker than countries of the West.

These findings make sense in view of then-prevailing availabilities of energy and capital. The Soviet Union has long produced surplus energy, which it provided (until 1982) at subsidized prices to its economic and political allies. By contrast, West Europe in the aggregate has been a heavy net importer of energy. Physical capital has been relatively more abundant in the West.

For East Europe, Moroney begins with economy-wide estimates of net real capital stocks, expressed in national currency units. He then uses purchasing power parity ratios (from the International Comparison Project) for each country to convert these stocks into 1975 U.S. dollars. These estimates are far from ideal. But they are the most nearly comparable international estimates of real capital thus far available. They form a basis for comparison with estimates of physical capital in West European countries, which Moroney develops from benchmarks published by Leamer.

Estimates of each country's aggregate energy consumption are taken from the United Nations, *Yearbook of World Energy Statistics*. Moroney then estimates coefficients of aggregate Cobb–Douglas production functions, where real GDP per worker depends on energy and capital per worker. The estimated coefficients of energy intensity in East Europe are roughly twice as large as those in West Europe. One may be uneasy about using such simple models as Cobb–Douglas now that a varied menu of flexible functional forms is available. It is a question of tastes. For trying to digest macroeconomic data subject to unknown margins of error, mine lean toward the simpler models. Anyway, this much is clear: Until at least the early 1980s, and quite probably well beyond, East Europe has been on average a far more intensive user of energy than West Europe.

What of the future? The Warsaw Pact has vanished, and with it Soviet energy subsidies and restrictive trade practices with East Europe. Germany stands politically unified once again, and within a few years will be economically integrated. Broad economic integration of West Europe will almost certainly be a reality by the end of 1992 or 1993. With financial capital from the West and Japan seeking international placement, one may expect faster real capital formation in the former Warsaw Pact nations. As these countries face rising real energy prices, their energy consumption patterns should more closely approximate those of West Europe. This evolution would be quickened if East Europe were to adopt binding environmental standards that reduce coal consumption. Liberalized trade coupled with freer capital and labor mobility will improve the lot of East Europe relative to the West.

ACKNOWLEDGMENTS

I am pleased to acknowledge numerous constructive comments on this introduction by M.A. Adelman, Danilo Santini, G. Campbell Watkins, and Peter Wilcoxen. I have included some, but not all, of their suggestions. I have almost certainly expressed some views here with which each of them may disagree.

FINDING AND DEVELOPING COSTS IN THE UNITED STATES, 1945–1986

M. A. Adelman

ABSTRACT

Development cost is defined as the ratio of development expenditures in a given year to reserves added in that year. Changes in development cost are a good proxy for changes in finding cost and in user cost, because discovery, development, and postponement or holding of hydrocarbons in place, are three competing forms of investment. Popular definitions of "finding" cost are an illogical and useless mixture of discovery and development. Although the discovery of large oil fields peaked before 1930, oil reserves added by development increased then stabilized around 1960. Costs tended if anything to decrease through 1972, but the decrease was mostly a one-time gain through the retreat from a costly regulatory scheme. The first price explosion in 1974 saw a strong *decline* in oil reserves added. The second price explosion was followed by an increase, but the best performance since 1949–51 came in 1983–85, when oil prices were declining by nearly one fourth in real terms. High oil and gas prices promoted a drilling boom, which raised factor prices and lowered efficiency. Old-field development was therefore in-

Energy, Growth, and the Environment
Advances in the Economics of Energy and Resources,
Volume 7, pages 11–58.
Copyright © 1992 by JAI Press Inc.
All rights of reproduction in any form reserved.
ISBN: 0-89232-947-5

hibited, but then helped as the boom deflated. Therefore the effect of the steeper price decline of 1986–87 has been mitigated by the decline in cost.

I. INTRODUCTION

The United States, excluding Alaska, is by far the largest and most intensively explored and developed oil province in the world. It is, therefore, the best place to study the effects of *diminishing returns over time* in oil and natural gas discovery and development.

Diminishing returns over time must be carefully distinguished from diminishing returns at any given time. The more wells to be drilled and reserves to be booked in a given year, the farther down the list of projects the industry goes. Moreover, haste makes waste. Therefore, under the conditions ruling at any given time, the greater the discovery–development effort, the less productive it is.

But over time, the largest fields would be found first even by chance, not to mention design; the better the drilling prospect, the earlier it is drilled. Hence over time there should be a persistent shift toward fewer and poorer reservoirs. The supply curve would move counterclockwise, all else being equal, and the price would rise (see Figure 5).

In the United States, by the end of 1945, 1.3 million wells had been dug, and 32 billion barrels produced (American Petroleum Institute [API], 1959). The industry was far down the discovery curve. Of the largest 208 fields in the "lower 48" (i.e., excluding Alaska) known in 1985, 120 had been found by 1930 and they contained over half of all the oil in the group (see Figure 1). In 1930 there were only 13 billion barrels left in "proved recoverable reserves" (American Petroleum Institute–American Gas Association [API–AGA], 1946). Yet through 1985, the United States, excluding Alaska, produce not 13 but 110 billion barrels.

It has been shown that reserves in known oil and gas fields continue to grow for decades after supposed maturity. Moreover, the growth of reserves in known fields in the United States after World War II was about equally divided between higher recovery rates and new oil in place (Adelman, Houghton, Kaufman, and Zimmerman, chap. 6). Proved reserves are only the ready shelf inventory of the industry, which keeps re-stocking the shelves by drawing from some undetermined amount "out there."

Diminishing returns have been extensively analyzed by estimating and projecting reserves and production (for recent surveys, see Meyer and Fleming, 1985; Woods, 1985). The best-known example of the physical approach is that of M. King Hubbert. He fitted a logistic curve to past production, and extrapolated it to predict future production, on the principle that the area under the ultimate curve was the original finite amount. After the peak, production would turn downward at a rate which would first accelerate, then flatten, converging toward zero.

Hubbert's prediction of 1970 as the peak production year was correct, apparently the only good prediction known to students of the oil industry. The objections of John M. Ryan, that the curve had no logical connection with the actual process of finding-developing-producing, seem never to have been much regarded (Hubbert, 1962; Ryan, 1965).

U.S. oil production in the "lower 48" declined slightly for a decade, but after mid–1980 remained quite steady. The industry seemed to take an unconscionably long time dying. Perhaps the higher prices were a reprieve, but there were also much higher costs. The much lower prices of 1986 are perceived as promising lower reserve-additions. The key is in the price-cost relation, which the volumetric approach ignores.

The problem can be posed by taking successive snapshots (*Oil & Gas Journal*, 1943, 1987) of two large oil fields (all amounts are in millions of barrels):

	Kern River (disc. 1899)	East Texas (disc. 1930)
End–1942 reserves:	54	2600
Cumulative production, 1943–1986	736	3031
End–1986 reserves:	970	1200

In more recent experience: the Prudhoe Bay field was rated for years at 9.6 billion barrels recoverable reserves. Early in 1987, it produced its fifth billion barrel, leaving, one might suppose, 4.6 billion. But this was becoming increasingly doubtful because the expected decline in output was postponed from year to year. In fact, an informed estimate shortly thereafter was of 8.2 billion, including 0.4 billion natural gas liquids (Salomon, 1987).

The additional barrels in large as in small fields were no gift of nature, nor did they reflect any "conservatism" in the original estimates. On the contrary, they were acquired by heavy investment both tangible and intangible. Our objective is to measure the relation between discovery–development investment and reserve-additions since World War II.

A. Some Costs and Pseudo-Costs

Many cost figures are mentioned these days, but usually the sources and methods are not explained. Often, there are obvious errors in one or both. For example, "finding costs" are often used to designate the sum of development and exploration outlays (Andersen, 1985). But this adds apples to oranges, and it compounds the error to compare the sum of the expenditures with the reserves described as "found" during the year. The reserves discovered through the finding effort of a given year will nearly all be booked in later years. As for money spent for the acquisition of acreage, that is not a cost at all, but a transfer payment.

Second, they add oil and gas, which have been subject to different forces, and reacted differently. This multiplies the effect of the first error. If Company A develops oil and Company B explores for gas, we add their expenditures, divide by Company A's reserves-added, and are alarmed at the average "finding cost." Each company knows better.

Third, the basic data are seriously biased downward. Thus, a compilation for 30 large companies, which accounts for about two-thirds of liquids reserves (Picchi and Winnall, 1986) has them replacing only 81 percent of their production in 1985, only 63 percent for 1978–1985 inclusive. (We exclude purchases.) But the corresponding Department of Energy totals for the whole industry were 134 and 112 percent.

These and other errors which we cannot trace cumulate into estimates of "finding cost" which are flights of fancy. The most notorious—though not the worst example—was the damages award in the Pennzoil–Texaco case. (We state no opinion on the legal question at issue, whether there existed a valid binding Getty–Pennzoil contract.) Pennzoil had paid about $3.40 per barrel for Getty's oil reserves. It claimed that the replacement cost by drilling would have been $10.87 (Petzinger, 1987). One need not believe that capital markets are perfect to see that such a 3:1 discrepancy between market value and replacement cost is ridiculous. Even more wild is an estimate submitted in October 1986 to the Independent Petroleum Association of America (IPAA) (which, we stress, was not their work) that "replacement cost of crude oil was over $26.50 in 1984, but if one adds financing costs the totals go much higher" (IPAA, 1986).

"Financing costs" are the cost of holding the asset oil-in-ground until it is extracted. For this reason, as we show below, oil at the wellhead has a cost or break-even price roughly three times as much as the capital cost

Finding and Developing Costs in the United States 15

or value of oil-in-ground. Hence this measure of so-called "finding cost" translates into a wellhead cost of about $80. In fact, even $26.50 as the in-ground cost is overstated by a factor of about six (see Table 6).

A recent article (Desprairies, Boy de la Tour, and Lacour, 1985) has some fairly elaborate cross classifications of reserves by cost category, but provides no hint of sources or methods. Moreover, a cost of $20 per barrel is said to be "compatible with a market cost of around $30/barrel" (p. 523), which sounds as though there is some additional undisclosed element. This mystery about the concept of cost makes it impossible to use.

Sometimes it is not even necessary to learn *how* an estimate was made to see that it is impossible. For example, there have been frequent references to an estimate that outside OPEC it takes $70,000 to find and develop one additional daily barrel of capacity (Franssen, 1985; Ebinger, 1985; Banks, 1985). This is presented as a worldwide parameter, to which the industry and its customers must adapt.

A little mental arithmetic shows this estimate to be impossible. In the United States, the cost of capital on equity funds is about 10 percent real, that is, assuming oil prices will move with the general price level. A rough average decline rate is around 12 percent (see Table 6). Assume 35 percent for royalties, state taxes (not income taxes), and operating costs (see Table 5). Spending $70,000 for a daily barrel only makes sense if the price is at least $65 per barrel, and is expected to rise with the rate of general inflation.[1]

During the delirium of 1979–1981 many oilmen expected such prices—some day. But it passes all credulity to suppose that they have on average been spending this, year in year out—without losing their shirts, their jobs, or their companies to takeovers or stockholders' suits.

In 1985, a barrel of developed reserves in the ground sold in the United States at $6 per barrel (*Oil & Gas Journal*, 1985c). Now, $6 per barrel in-ground equates to $20,300 per initial daily barrel of capacity (see Equation [2]). Rational people will not spend the equivalent of $70,000 for what they can reproduce for less than one-third the amount.

The $70,000 per daily barrel delusion is a useful reminder. A cost estimate needs to be validated by reference to the relevant price. If it passes the test, it may still be wrong, but if costs are far above prices, the estimate must be rejected, and the estimator must go back to the drawing board.

II. THEORY

Costs are measured as an investment outlay versus (1) new reserves in the ground, or (2) new productive capacity. There is a basic relationship between (1) and (2). Proved reserves are the amount which will ultimately be produced out of a pool by the capacity of facilities in place. Hence if R = reserves, Q = initial output, and a = the exponential decline rate, then:

$$R = Q e^{-at} \int_0^T dt = Q/a\,(1 - e^{-aT}) \quad (1a)$$

As T becomes large, R approaches Q/a, or $a = Q/R$. (1)

With normal pool lifetimes, the error in using infinite time is usually but not always negligible. The depletion rate is only an approximation to the true decline rate, and is subject to biases up and down (for a fuller discussion, see Adelman et al., 1983, Appendix B)

In the United States, good data exist on annual increments to proved reserves, and development costs can be calculated as dollars per barrel added in the ground. But with Equation (1), that figure can be translated into outlays per initial barrel of capacity, and checked against independent data.

For example, if K = investment, and K/R, the cost of installing facilities which will enable us to book one barrel in the ground, then the investment per additional daily barrel in the United States in recent years, when the decline rate a was about 0.12:

$(T \to \infty)$ $K/Q = 365\,(K/Ra) = 365K/.12 = 3042\,K$

$(T = 25)$ $K/Q = 365\,(K/Ra)(1 - e^{-aT}) = 365K/.113 = 3230K$ (2)

Conversely, if we learn the investment per daily barrel is, for example, $20,000 per daily barrel, the investment per barrel in-ground is about $20,000/3230K = $6.19.

Moreover, as indicated earlier, the producer needs to hold the asset, as a stock of proved reserves, until he sells it off. Thus the real supply price must allow for the ratio of above-ground to in-ground values.

Defining K, a, R, T, and Q as before, we add i, the minimum acceptable rate of return, and P as the market value.

Undiscounted value of in-ground reserves = $PR = PQ/a$

Discounted present value above ground =

$$VR = \int_{t=0}^{T} PQ\, e^{-(a+i)t}\, dt = PQ/(a+i) = PR\, a/(a+1) \qquad (3)$$

Then in equilibrium the *value* of a discounted above-ground unit relative to an undiscounted below-ground unit is approximately $(a+i)/a$. What comes to the same thing, the *cost* of holding the inventory below-ground until the time of production is

$$(T \to \infty) \quad (a+i)/a = 1 + (i/a) \qquad (4)$$

$$(T = 25) \quad (a+i)/a = (a+i)/a(1 - e^{-(a+1)T}) \qquad (4a)$$

Assuming $R/Q = 12$, $i = 0.1$, the cost of holding is 1.833 assuming infinite time, and 1.783 assuming 25 years.

Thus a barrel or *mcf* above ground is barely worth buying at a price which is $(1 + (i/a))$ times the price of a unit below ground. Contrariwise, if a unit above ground cannot be expected to sell for $(1 + (i/a))$ times its cost to create under ground, then it is not worth creating.

Obviously, the higher the discount rate, and the longer it takes to get the oil or gas out of the ground (reciprocal of a), the more expensive is the oil or gas. The faster the depletion, and the shorter the holding time, the better—all else being equal.

But faster depletion takes more investment. Hence the optimum depletion rate is a tradeoff between higher investment and quicker return. (Of course, the depletion rate may be limited by government, or by some kind of monopoly arrangement limiting investment and production.)

But, instead of depleting a pool faster and more expensively to get more production, at some point it pays to incur the costs of *finding* an additional pool. Thus development and exploration are limited substitutes for each other. This hint is followed up later.

III. DATA: EXPENDITURES AND RESERVE ADDITIONS

Column 1 of Table 1 presents a series for converting nominal into constant-dollar expenditures.[2] In columns 2–7, using column 1 as deflator, we show expenditures made for finding and for developing, 1955–1986. The *Joint Association Survey* is the source for 1955–1972, the Census Bureau for 1973–1982. The estimates for 1983–1986 are approximations based on the J.A.S. drilling expenditures, since total exploration or total development are no longer available from any source.[3] The expenditure series have been purged of lease bonuses or lease rentals, which are not costs, but rather transfer payments, that is, a share of past or expected profits, paid to the landowners, chiefly the U.S. government. The division of exploration expenditures between oil and gas is proportional to the number of successful exploratory wells for each; total development expenditures are divided between oil and gas in proportion to drilling expenditures respectively on development oil wells and development gas wells.

In Table 2, the deflated expenditures of Table 1 are divided by reserves-added, to obtain the cost per unit added. The reserve additions were published for many years by the American Petroleum Institute and American Gas Association (API-AGA, 1959–1979; Energy Information Administration [EIA], 1977–1986). They are understated because of the omission of natural gas liquids, which, following the EIA takeover of the reserve statistics, are no longer compiled by origin (EIA, personal communication, December 5, 1985). It is an unfortunate gap, but with little effect on the observed trend.

Increments to gas reserves include only nonassociated gas, since we are trying to match them with drilling and equipping and other expenditures on new gas wells. This again gives some understatement because some part of the expenditures for crude oil development is for the production of associated-dissolved gas. An alternative series shown in Table 4 and Figure 6 is essentially a combined finding-developing cost per barrel.

Table 3 shows some factors bearing upon cost changes: the depletion rate, cumulative production, number of development wells drilled, and total drilling expenditures per rig year, a rough indicator of efficiency. The increase in drilling expenditures per rig year appears to be an anomaly, given the efficiency increases that have occurred in the industry since the post–1979 drilling boom. Certainly the increase in the number of wells drilled per rig year acts to offset the rise in expenditures per rig year, but a decrease in expenditures would still have been expected, or at least stability, not an increase. The explanation for this result is derived, at least in part, from the fall in the drilling price index (Table 1) which offsets the drop in nominal drilling expenditures.

It might at first appear that these reserve and expenditure data were so highly aggregated as to be useless. After all, they include a very large number of fields and reservoirs, and a wide range of recorded costs. And

Table 1. Discovery and Development Expenditures, 1955–1986
(In millions 1983 dollars)

	(1)	(2)	(3)	(4)	(5)	(6)	(7)
	Drilling		Exploration			Development	
	Cost Index		Crude	Nonassociated		Crude	Nonassociated
Year	(1983 = 100)	Total	Oil	Gas	Total	Oil	Gas
1955	21.4	7150	5141	2009	11679	10682	997
1956	22.9	7695	5647	2048	11796	10432	1364
1957	24.2	n.a.	n.a.	n.a.	n.a.	n.a.	n.a.
1958	24.2	n.a.	n.a.	n.a.	n.a.	n.a.	n.a.
1959	24.6	5888	4250	1638	10378	7492	2887
1960	24.7	5762	3878	1885	9322	6273	3049
1961	24.5	5924	3965	1959	9389	6284	3105
1962	24.7	6175	4144	2031	10182	6833	3349
1963	24.9	5936	4202	1734	9120	6456	2664
1964	25.2	6246	4221	2025	9595	6484	3111
1965	26.3	5984	4111	1873	9000	6184	2817
1966	27.0	5942	3832	2110	9230	5952	3278
1967	28.0	5825	3872	1954	8947	5947	3000
1968	29.5	5649	3886	1763	8591	5910	2681
1969	30.9	5936	3848	2088	8948	5801	3147
1970	32.8	4946	3155	1792	8684	5538	3146
1971	35.8	4491	2665	1825	7459	4427	3032
1972	38.0	4756	2645	2111	8136	4525	3611
1973	41.7	4946	2047	2899	8214	4629	3585
1974	51.0	6319	2628	3691	9303	5101	4202
1975	60.1	6991	2499	4492	11993	6781	5212
1976	65.9	7105	2235	4870	13847	6869	6978
1977	73.8	7740	2651	5089	13410	6592	6817
1978	82.8	8837	2905	5932	15059	6535	8524
1979	96.0	12375	4076	8299	16918	7453	9465
1980	110.9	15868	5923	9944	19744	9329	10415
1981	130.8	20739	8502	12237	24473	12621	11852
1982	138.5	20004	7296	12707	24133	11263	12870
1983	100.0	15983	5324	10659	22198	13043	9154
1984	95.9	16521	7038	9483	23384	14192	9193
1985	95.9	15560	6832	8729	21953	12854	9099
1986	84.9	10459	3987	6473	13962	8452	5510

Sources:

Column 1: Drilling Cost Index (DCI)

1955–1962: Implicit Price Deflator for Total Gross Private Non-Residential Investment (IPDNRT) used from the *Economic Report of the President*. DCI (i) = DCI63/IPDNRT63) IPDNRT (i); i = year. The implicit price deflator and the drilling cost index correlelate closely for 1963–1972.

1963–1985: Independent Petroleum Association of America, *Report of Cost Study Committee*, cost index refers to drilling and equipping wells.

1986: IPAA, change in judgmental index from May 1987 *Report of the Cost Study Committee*.

Columns 2–7: Expenditures (see detailed calculation notes below)

Note: The sources are for Expenditures in Current Dollars. Expenditures in Constant Dollars were calculated using the Drilling Cost Index.

1955–1972: *Joint Association Survey* (Section II).

(continued)

Table 1. (Continued)

1973–1982: *Annual Survey of Oil and Gas*, U.S. Bureau of the Census. Most numbers used in the calculations are from tables giving data on "Net Company Interest Statistics" (Table 3 in earlier issues and Table 5 in later issues). Gross numbers are from earlier issues, Table 4 in later issues).

1983–1986: See below.

Column 2:

Exploration Total

1955–1972: (total exploration expenditures) minus (lease acquisition plus lease rental) plus (exploration overhead)

1973–1982: ((total exploration expenditures) minus (lease acquisition plus lease rental)) * ((gross expenditures drilling and equipping exploration wells)/(net expenditures drilling and equipping exploration wells))

1983–1986: prior year's total exploration expenditures * ratio of current to prior years' total exploratory drilling expenditures. (*Joint Association Survey*)

Column 3: Exploration Oil

1955–1956: (column 2) * (no. of oil wells (development plus exploration)) ((no. of oil plus gas wells (development plus exploration))

1959–1972: (column 2) * (total oil well expenditures (exploration and development))/(total oil plus gas expenditures (exploration and development))

1973–1982: (column 2) * (gross expenditure drilling and equipping exploration oil wells)/(gross expenditure drilling and equipping exploration oil plus gas wells)

1983–1986: (column 2) * (exploratory drilling expenditures minus oil wells/exploratory drilling expenditures minus oil and gas wells) (*Joint Association Survey*)

Column 4: Exploration Nonassociated Gas

Column 2 minus column 3.

Column 5: Development Total

1955–1972: (total development expenditures) plus ((development plus production overhead) * (total development expenditures)/(total development plus production expenditures))

1973–1982: ((total development expenditures) minus (development acquisitions)) * ((gross expenditures drilling and equipping development wells)/(net expenditures drilling and equipping development wells))

1983–1986: prior years total development expenditures * ratio of current to prior years' total development drilling expenditures (*Joint Assocaition Survey*)

Column 6: Development Oil

1955–1956: (column 5) * (no. of oil wells (development plus exploration))/(No. of oil + gas wells (development plus exploration))

1959–1972:)(column 5) * (total oil expenditures (exploration and development))/(total oil plus gas expenditures (exploration and development)

1973–1982: (column 5) * (gross expenditure drilling and equipping development oil wells/gross expenditure drilling and equipping developnment oil plus gas wells)

1983–1986: (column 5) * (development drilling expenditures minus oil wells/development drilling expenditures minus bil and gas wells)

Column 7: Development Nonassociated Gas

Column 5: minus column 6.

it is true that if these data aggregated the *average* or *total lifetime* cost of many reservoirs, the result would not be interesting. But a single year's data record *incremental* (not average) cost across all reservoirs developed. The industry is a selective mechanism for maintaining or expanding output at the least cost. Under competitive conditions (which hold in the United States, though not in the world industry), marginal cost everywhere moves toward the expected price. Lower-cost reservoirs are expanded to the point where rising costs choke off additional drilling. Higher-cost reservoirs are drilled more selectively, to bring down in-

Table 2. Reserves and Unit Costs

Year	Reserves Added Crude Oil (millions barrels) (1)	Reserves Added Natural Gas: Nonassociated (bcf) (2)	Unit Costs (1983 dollars) Crude Oil ($/bbl) (3)[a]	Unit Costs (1983 dollars) Natural Gas: Nonassociated ($/Mcf) (4)[b]
1947	2465	6464	n.a.	n.a.
1948	3795	7483	n.a.	n.a.
1949	3188	6845	n.a.	n.a.
1950	2563	9059	n.a.	n.a.
1951	4414	8475	n.a.	n.a.
1952	2749	9615	n.a.	n.a.
1953	3296	15363	n.a.	n.a.
1954	2873	5627	n.a.	n.a.
1955	2871	12748	3.72	0.08
1956	2974	15297	3.51	0.09
1957	2425	16205	n.a.	n.a.
1958	2608	17382	n.a.	n.a.
1959	3667	14782	2.04	0.20
1960	2365	11545	2.65	0.26
1961	2658	15147	2.36	0.20
1962	2181	18017	3.13	0.19
1963	2174	12256	2.97	0.22
1964	2665	17366	2.43	0.18
1965	3048	18431	2.03	0.15
1966	2964	17037	2.01	0.19
1967	2962	17964	2.01	0.17
1968	2455	13979	2.41	0.19
1969	2120	6855	2.74	0.46
1970	2890	9340	1.92	0.34
1971	2318	8917	1.91	0.34
1972	1558	7812	2.90	0.46
1973	2146	3988	2.16	0.90
1974	1994	7037	2.56	0.60
1975	1318	10371	5.14	0.50
1976	1085	7460	6.33	0.94
1977	1272	9451	5.18	0.72
1978	1965	14774	3.33	0.58
1979	1808	14069	4.12	0.67
1980	2970	11434	3.14	0.91
1981	2570	19877	4.91	0.60
1982	1382	14328	8.15	0.90
1983	2897	13415	4.50	0.68
1984	3748	12508	3.79	0.73
1985	3022	9593	4.25	0.95
1986	1446	12895	5.85	0.43

Notes: *Column 6, Table 1/Column 1.
**Column 7, Table 1/Column 2.

Sources:

Columns 1,2: Reserve Additions by Development.

Oil: 1946–1976: *API/AGA* (Table 2, adjusted for Alaska [1970] using Table 3-2).

1977–1979: Average of API/AGA data and *Energy Information Administration Annual*, U.S. Energy Information Administration Reserves.

1980–1986: *Energy Information Administration*.

Gas: 1945–1959: Adelman (1962, Table IV-B).

1960–1965: Adelman (1962).

1966–1977: *API/AGA* (Table VIII).

1978–1979: Average of *API/AGA* and *EIA* figures.

1980–1986: *EIA*.

Note: Tables references are from sources given.

Note on Alaskan exclusion: Oil: Only reserves discovered in 1968 and developed in 1970 were excluded. Gas: Backdated discoveries inlucde associated and nonassociated gas; therefore, Alaska excluded in 1968. Developed reserve additions include only nonassociated gas; therefore, no Alaskan exclusion made as nonassociated volumes are insignificant.

Table 3. Factors Bearing on Cost Changes

	(1)	(2)	(3)	(4)	(5)	(6)	(7)
	\multicolumn{2}{c}{Production/Reserves (Excluding Alaska)(percent)}	\multicolumn{2}{c}{Cumulative Produciton (Excluding-Alaska)}	\multicolumn{2}{c}{Development Wells Drilled}	Total Drilling Expenditures
Year	Crude Oil	Nonassociated Gas	Crude Oil (mil. bbls)	Gas (mil boe)	Crude Oil (000s)	Gas (000s)	Per Rig Year (million 1983 dollars)
1947	8.61	2.63	35024	2152	17.02	2.91	n.a.
1948	8.60	3.11	37026	2791	21.49	2.53	n.a.
1949	7.38	3.27	38845	3478	20.64	2.46	n.a.
1950	7.69	3.48	40789	4237	22.85	2.41	n.a.
1951	8.06	3.99	43003	5125	21.69	2.58	n.a.
1952	8.07	4.17	45260	6083	21.69	2.70	n.a.
1953	7.99	4.19	47572	7108	23.78	3.11	n.a.
1954	7.64	4.38	49829	8175	27.79	3.25	n.a.
1955	8.06	4.45	52248	9303	29.33	2.74	7.01
1956	8.39	4.57	54800	10523	28.46	3.72	7.44
1957	8.45	4.61	57359	11817	26.07	3.76	n.a.
1958	7.77	4.49	59731	13150	22.83	3.98	n.a.
1959	7.83	4.60	62214	14562	24.10	4.12	7.84
1960	7.82	5.05	64684	16130	19.86	4.39	8.62
1961	7.91	5.10	67190	17753	19.94	4.85	8.70
1962	8.11	5.02	69730	19413	20.04	5.08	9.99
1963	8.36	5.41	72312	21214	18.97	4.09	10.04
1964	8.52	5.58	74945	23125	19.40	4.28	10.55
1965	8.58	5.79	77620	25164	17.82	4.21	10.80
1966	9.15	6.02	80470	27319	15.42	3.59	11.92
1967	9.71	6.25	83479	29592	14.34	3.13	13.03
1968	10.08	6.82	86537	32049	13.38	2.97	12.18
1969	10.69	7.72	89658	34711	13.28	3.47	12.47
1970	11.21	8.59	92893	37563	12.28	3.35	13.26
1971	11.37	8.93	96070	40407	11.24	3.51	12.24
1972	12.22	9.77	99278	43356	10.75	4.83	11.65
1973	12.36	10.67	102391	46325	9.60	5.90	11.01
1974	12.30	10.90	105363	49173	12.79	5.96	10.62
1975	12.44	10.44	108179	51803	15.99	6.91	11.44
1976	13.06	10.91	110941	54404	16.60	8.08	12.65
1977	12.39	10.98	113636	56957	17.52	10.56	10.57
1978	12.56	11.14	116181	59531	17.88	12.62	10.60
1979	12.52	11.98	118626	62257	19.35	13.25	13.46
1980	11.32	11.43	121010	64912	30.34	15.05	12.24
1981	11.15	11.19	123367	67589	39.86	17.22	11.39
1982	11.36	10.39	125690	70065	35.61	16.24	14.21
1983	11.53	9.09	128045	72238	33.68	12.33	17.13
1984	11.49	10.14	130444	74626	37.06	13.83	16.44
1985	11.17	9.49	132829	76816	35.02	11.74	19.05
1986	11.51	9.19	135132	78946	18.67	7.65	25.33

Sources:

Columns 1,2: Production/Reserves
Oil: Total United States: 1947–1976; *API/AGA* (Table II).
  1977–1979: Average of *API/AGA and EIA* figures.
  1980–1986: *EIA*.
Alaska: 1947–1958: N.A., Volumes are insignificant.
  1959–1976: *API/AGA* (Table III-2).
  1977–1979: Average of *API/AGA* and *EIA* figures.
  1980–1986: *EIA*.

(*continued*)

Table 3. (Continued)

Non-associated gas: Total United States
    1947–1959: Adelman, *The Supply and Price of Natural Gas* (Table IV-B).
    1960–1965: Adelman methodology; see above.
    1966–1976: *API/AGA* (Table VIII).
    1977–1979: Average of *API/AGA* and *EIA* (figures).
    1980–1986: *EIA*.

Alaska:
  reserves: 1947–1958: N.A. = Volumes are insignificant.
          1980–1986: *EIA*.
  production: 1947–1958: N.A. = Volumes are insignificant.
             1959–1976: *API/AGA* (Volume 32, Table XII-3).
             1977–1978: Average of *API/AGA* and *EIA* figures.
             1979–1986: *EIA*.

Mcf converted to boe by ratio (1 boe/6.0 Mcf). Reserves are year-end
Note: Tables referenced are from sources given.
Columns 5,6: American Association of Petroleum Geologists, Annual North American Drilling issues, later merged with World Energy Development issues; and *Oil & Gas Journal*.
Column 7: (Columns 2 and 5 from Table 1)/(number of rig years). The number of rig years, from Hughes Tools Co., Reported in *Oil & Gas Journal*, *World Oil*, *Basic Petroleum Data Book*, and so forth.

cremental costs, or are not drilled at all. The industry is forever approaching the long-run equilibrium, which of course it never attains. (The excluded transfer payments equalize private marginal costs, not social costs.)

Of course, there is still a great variability among new projects, some real and some artificial. On the low side: new low-cost reservoirs are drilled up only gradually. Costs are abnormally low at first, then rise as they approach marginal equality. This biases the total down. Contrariwise, the early stages of any project are outlays with nothing to show for them, which biases the total expenditure up.

Indivisibilities are also a distortion. Incremental cost in a pool may be below the price, but more intensive development would require so large an additional investment, such as another platform, that its cost would exceed the price. More generally, too high a production rate will damage the reservoir and exact a very high cost; best not to approach the edge of the cliff.

So far, reserves *developed* during the year have been discussed. Oil in newly found fields and pools can only be roughly estimated, as the American Association of Petroleum Groups (AAPG) does annually ("World Energy Developments"). For those years where they overlap, the AAPG estimates are below those of the API, and I would consider them downward-biased (Meyer and Fleming, 1985, p. 1953). Back-dated estimates made years later are much more accurate.

No use is made here of the item "discoveries" in the API–AGA or the current EIA publications. These "discoveries" include only that small part of newly-found fields that has actually been booked, that is, developed and made ready for production in the given year. The eventual reserve figure is several or many times the initial estimate, with great variation among fields, regions, and years. Hence the annual "discoveries" number is meaningless, a fragment masquerading as data.

For the Province of Alberta in Canada, there are back-dated discoveries, year by year, and by fields, since 1947 (Alberta, 1984; Uhler and Eglington, 1986). For the United States, similar numbers by states (but not fields) are available from 1921 through 1979. After that year, the API–AGA statistics were replaced by a Department of Energy series. Unfortunately, this was based on a sample of operators not fields, and the back-dated estimates could no longer be made. And because it takes about six years to get a reliable estimate, the usable back-dated entries end in 1973. A panel appointed by the National Academy of Sciences (NAS) has criticized this procedure (NAS, 1985), and pointed out the statistical gap.

Because fields continued to grow after 1973, even the API-AGA series is too low. Hence we cannot use it to calculate unit costs. Even if we did, since a biased series might be better than none, a given year's "cost" would not be comparable with, nor additive to, the same year's unit development cost. The next section shows how changes in development cost are a proxy for changes in discovery costs; and how discovery values may be estimated from reserve values and development costs.

IV. DISCUSSION: GRAPHIC SUMMARIES

A. Oil and Gas Discovery

Figure 1 is based on the original reserves credited as of the end of 1985 to 186 large fields which account for approximately 50 percent of total U.S. reserves. The lower line shows the percentage credited to fields found during the decade ending in the year shown. Thus fields found before 1900 contained about 5 percent of the total, 1901–1910 discoveries about 7 percent, and so forth. Oil discoveries peak in the decade 1921–1930, with about one-fourth of the total, after which the decline is rapid to 1960 and thereafter. The upper curve cumulates the decades. It

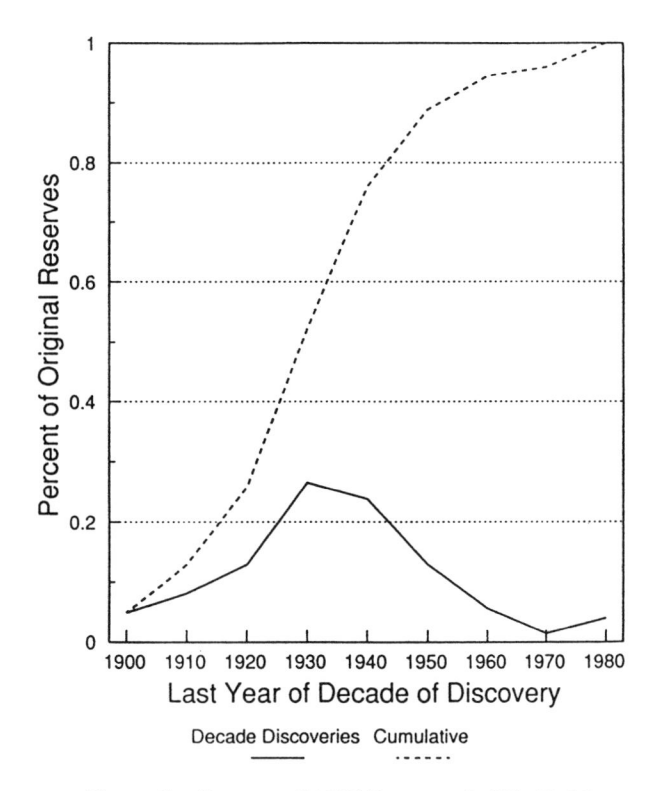

Finding and Developing Costs in the United States 25

Figure 1. Percent of 1985 Reserves in Big Fields
(5MB reserves of more, excluding Alaska)

shows, for example, that about 80 percent of the oil in these large fields discovered through 1985 was in fields discovered before 1945.

Figure 2 shows the trends in backdated discoveries of oil and gas in all fields, large and small, during 1920–1973. The oil line checks approximately with the trends shown in Figure 1, with the same peak before 1930, and the severe decline since. The AAPG series (not shown) at least suffices to show there was no reversal of the decline after 1973. Nonassociated gas discoveries are dominated by the spike in 1922; another much smaller peak is in 1951.

Figures 3 and 4 show 1955–86 expenditures on exploration and development for oil and gas in current dollars and constant dollars, respectively.[4] In oil, there was a slow decrease in real expenditures through 1973, while gas shows a drop in the later fifties, and a rough constancy afterward. Then comes the big boom. Stated in real terms, oil

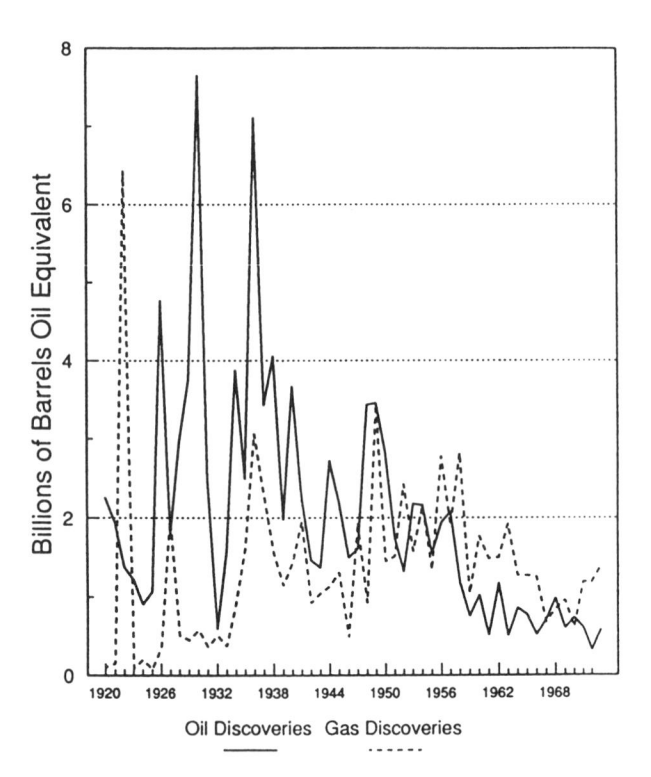

Figure 2. Oil and Gas Discoveries
(Excluding Alaska)

development expenditures decline sharply after 1955–1956, while the other classes remain stable.

B. Oil Development Costs

As stated earlier, diminishing returns in finding new fields and pools are to be expected. Development investment would at first seem to be an activity of gradually increasing efficiency because of progress in technology. But this is only true ceteris paribus, when considering a given operation, for example, drilling to a given depth in a given place. It is not true of the observed development cost at any time.

If newer fields are getting smaller, deeper, more heterogeneous and faulted, and so forth, then development cost per unit of reserves booked

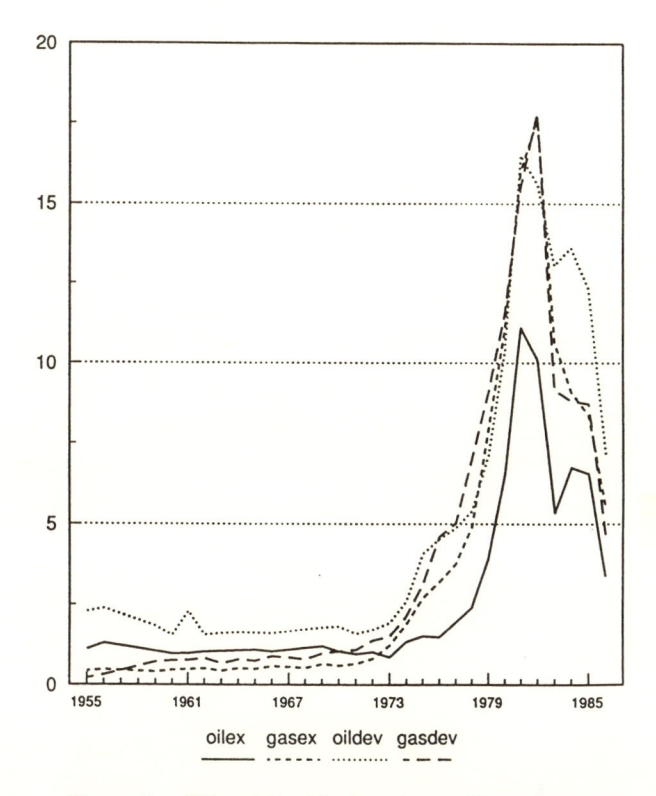

Finding and Developing Costs in the United States 27

Figure 3. Oil and Gas Exploration and Development: Expenditures in Billions Current Dollars

in those new fields must also increase.[5] Moreover, when it becomes increasingly difficult to find new oil fields as good as those previously discovered, the alternative of more intensive development in the old fields becomes more attractive. Thus more development wells will be drilled into and near the older pools.

Finally, the higher the cost of new oil, the greater the incentive to drain the old oil faster, even at the cost of higher investment requirements. As already discussed, the higher expenditures are to an important degree offset by the faster payout. Of course, in each reservoir, a point is reached where faster depletion would be inordinately costly because it would damage the reservoir.

Thus for three (not wholly independent) reasons, oil development cost changes reflect discovery cost changes: as the crop gets more scanty,

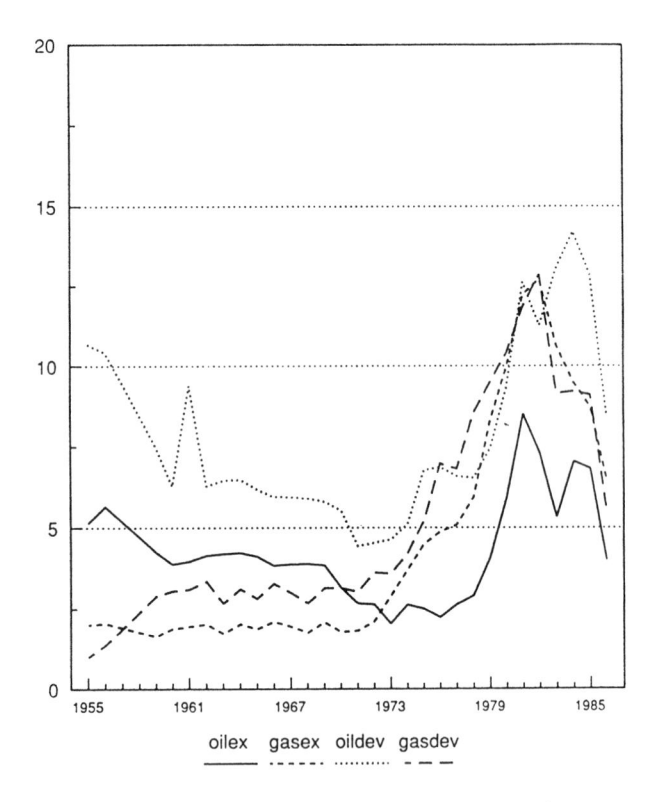

Figure 4. Oil and Gas Exploration and Development:
Expenditures in Billions 1983 Dollars

more development effort is needed to process it. Hence oil development costs are a proxy or indirect indicator of changes in finding cost.

The general principle is, at the margin, another unit of investment ought to bring the same return in development as in exploration. With lower development cost ceteris paribus, the shift toward development from exploration relieves the pressure on exploration, and stops when marginal returns are again equated.

It is natural to think of discovery and development as complementary, and so they are in any given project. The higher the expected finding cost, the lower must the development cost be for the project to be acceptable. But outside of a relatively small number of projects in the early stages, discovery and development are overwhelmingly substitutes not complements.

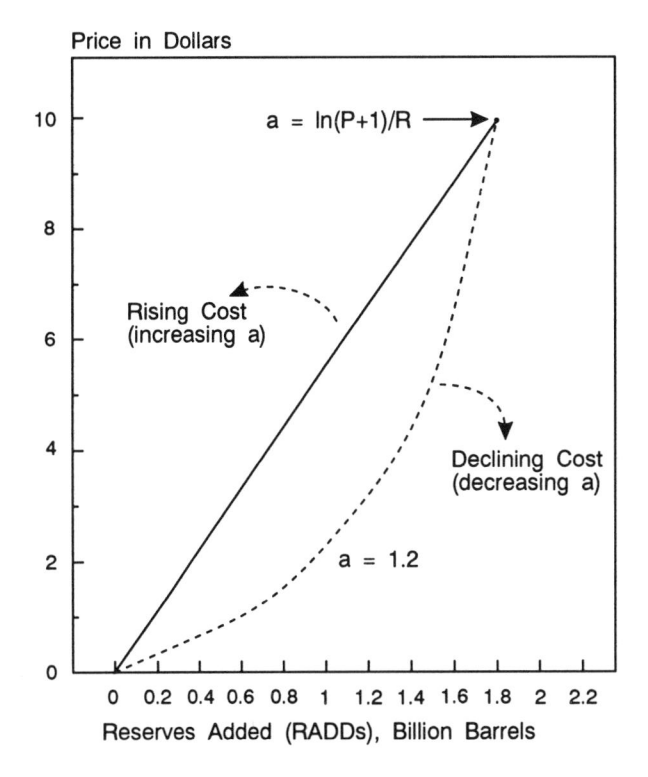

Finding and Developing Costs in the United States 29

Price in Dollars

Figure 5. Hypothetical Supply Curves: Corresponding to a Given Price, RADDs

C. A Measure of Changes in Finding-Developing-Producing Cost

The unit costs calculated above, dividing expenditures by reserves-added, are deficient in that they must to some extent reflect movements along the supply curve, when we are trying to isolate the movement of the curve. In Figure 5, reserves-added are plotted on the horizontal axis, price on the vertical. We have one data point, which represents the intersection of the year's supply curve with the year's demand curve. The curve must pass through the origin, since with a zero price there would be zero reserve-additions. The curve cannot possibly lie to the left of the straight line from the origin, and it is almost surely concave upward. We approximate it by a logarithmic curve. Thus if 2 billion barrels were added at a price of $10, the slope coefficient would be 1.2.

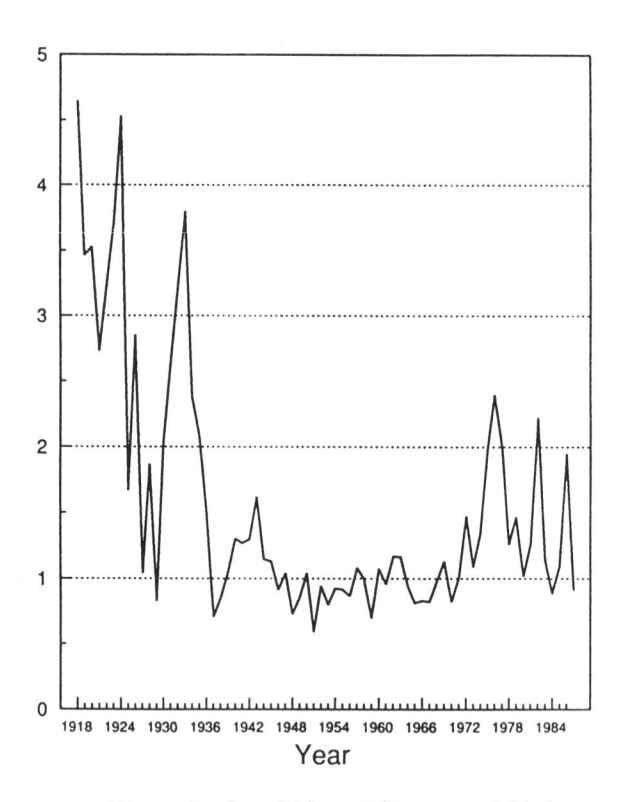

Figure 6. Log (Price +1)/Reserves Added: 1918–1987 (1983 $)

This is of course an ordinal measure only. It includes discovery or inground value, development cost, and current producing cost. Over time, an increasing slope of the curve would signal rising costs, even if we knew nothing about the slope of any particular curve.

Table 4 and Figure 6 show the slope of the price: reserves-added relationship over a 70-year period. During the discovery boom 1918 to 1929, the slope decreased by a factor of about 4, despite much fluctuation. Omitting the depression outliers of 1930–1937, when investment dwindled and reserves-added went nearly to zero, a small net decrease is apparent. For the next 36 years, when exploration dwindled rapidly, the supply curve showed no leftward rotation.

A similar price/reserves plot for natural gas is not possible, because the published "price" of natural gas is merely the arithmetic average of

Table 4. Prices and Reserve Additions, 1918–1986

Year	Current Price	Price (1983 $)	Reserve Additions (MM bbl)	Price/Reserve Additions (1983 $/ bil. bbl)
1918	1.98	20.02	656	30.52
1919	2.01	19.89	878	22.66
1920	3.07	26.79	943	28.41
1921	1.73	17.64	1072	16.45
1922	1.61	18.29	358	51.08
1923	1.34	14.14	732	19.32
1924	1.43	15.16	614	24.69
1925	1.68	17.99	1764	10.20
1926	1.88	20.21	1071	18.87
1927	1.30	14.05	2601	5.40
1928	1.17	12.62	1401	9.01
1929	1.27	13.28	3207	4.14
1930	1.19	12.92	1298	9.95
1931	0.65	7.62	251	30.36
1932	0.87	11.35	85	133.47
1933	0.67	9.00	606	14.86
1934	1.00	12.21	1085	11.25
1935	0.97	11.64	1220	9.54
1936	1.09	12.97	1763	7.36
1937	1.18	12.86	3722	3.45
1938	1.13	12.19	3054	3.99
1939	1.02	11.14	2399	4.64
1940	1.02	10.77	1893	5.69
1941	1.14	11.17	1969	5.67
1942	1.19	10.53	1879	5.60
1943	1.20	10.03	1484	6.76
1944	1.21	9.75	2068	4.71
1945	1.22	9.76	2100	4.65
1946	1.41	10.35	2658	3.89
1947	1.93	12.08	2465	4.90
1948	2.60	14.82	3795	3.91
1949	2.54	14.07	3188	4.41
1950	2.51	13.52	2563	5.28
1951	2.53	12.55	4414	2.84
1952	2.53	12.36	2749	4.50
1953	2.68	12.86	3296	3.90
1954	2.78	13.23	2873	4.60
1955	2.77	12.93	2871	4.50
1956	2.79	12.17	2974	4.09
1957	3.09	12.78	2425	5.27
1958	3.01	12.46	2608	4.78
1959	2.90	11.79	3667	3.22
1960	2.88	11.66	2365	4.93
1961	2.89	11.78	2658	4.43
1962	2.90	11.74	2181	5.38
1963	2.89	11.62	2174	5.35
1964	2.88	11.41	2665	4.28
1965	2.86	10.87	3048	3.57
1966	2.88	10.68	2964	3.60
1967	2.92	10.42	2962	3.52
1968	2.94	9.98	2455	4.06
1969	3.09	10.00	2120	4.72
1970	3.18	9.69	2890	3.35
1971	3.39	9.47	2318	4.08
1972	3.39	8.92	1558	5.72
1973	3.89	9.34	2146	4.35
1974	6.87	13.48	1994	6.76

(continued)

Table 4. (Continued)

Year	Current Price	Price (1983 $)	Reserve Additions (MM bbl)	Price/Reserve Additions (1983 $/ bil. bbl)
1975	7.67	12.76	1318	9.68
1976	8.19	12.44	1085	11.46
1977	8.57	11.61	1272	9.13
1978	9.00	10.88	1965	5.53
1979	12.64	13.17	1808	7.28
1980	21.59	19.47	2970	6.56
1981	31.77	24.30	2570	9.45
1982	28.52	20.59	1382	14.90
1983	26.19	26.18	2897	9.04
1984	25.88	27.00	3748	7.20
1985	24.09	25.85	3022	8.55
1986	12.51	13.79	1446	9.53

Sources:

Current price:

1918–1944: Calculated from *Crude Price Index, Twentieth Century Petroleum Statistics* (1986), Degolyer and MacNaughton (p. 98).

1945–1945: U.S. wellhead price from *EIA Annual Energy Review*.

Deflator:

1918–1945: *Long Term Economic Growth 1860–1965* (U.S. Department of Commerce, October 1966, p. 200).

1945–1986: Drilling Cost INdex from Table 1 used.

Reserve addtion:

1918–1944: See source for Current Price, Col. 1.

1945–1986: See sources for Table 2.

all prices on old and new contracts, interstate and intrastate, in no way comparable with current costs, hence with no meaning for supply. Of course they were in any case distorted during the long period when gas prices were under control—as, indeed, many still are—and were forced artificially high or low.

Figure 7 shows, for the period since 1955, both measures of the cost of oil, together with the development cost of gas. The two oil measures move generally together, but the price of gas shows little relation, until of course the 1970s.

D. Development Costs 1973–1985

There was a sharp break with the past after 1972, and both development costs and total costs increased very greatly, for various reasons. Wells drilled rose by a factor of 3.25. The demand for oil and gas drilling-equipping services greatly exceeded supply, which could not be quickly expanded. Hence the price of drilling equipment services rose very sharply. But even adjusted for factor price changes, the real cost

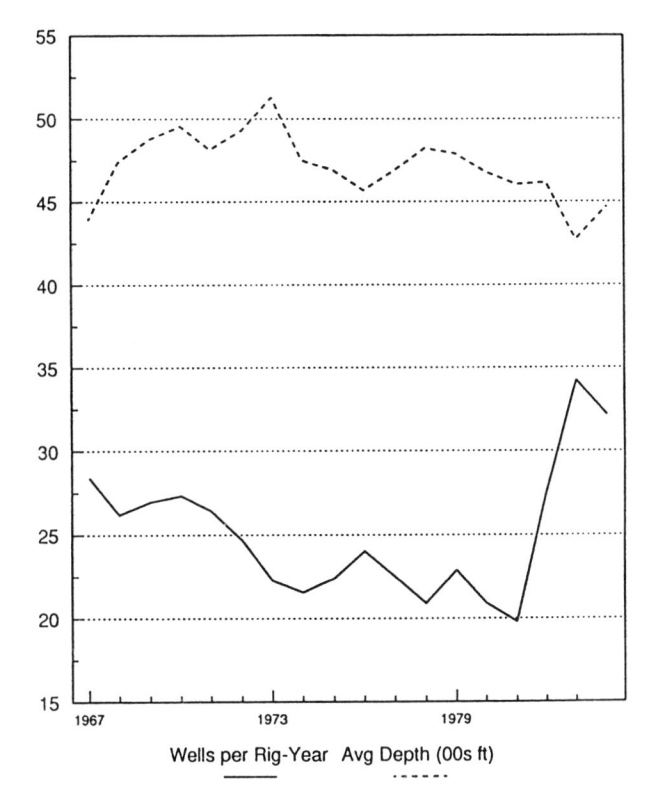

Finding and Developing Costs in the United States 33

Figure 7. Wells (All Types) per Rig-Year

approximately doubled over that time, whether measured by outlays per reserve barrel added, or by the price of oil related to reserve barrels.

Obviously, the services were used much less efficiently. Anecdotal evidence abounds. It is all too credible that kickbacks inflated drilling costs 30 to 40 percent in some instances (*Wall Street Journal,* 1985). But my own conjecture is that the use of untrained personnel, and the hoarding of men, materials, and machines, were much more important. In any case, the return to drilling investment had to be less, as the industry moved up the short-run supply curve.

Figure 7 shows that during 1949–1968, although well depth increased, the number of wells drilled per rig year persistently increased, showing increasing efficiency. During the next five years, well depth continued to increase, and wells per rig year decreased, indicating perhaps unchanged efficiency. But after 1973, although average depth decreased, wells drilled per rig year continued to decline, indicating a loss of efficiency, until the startling reversal after 1981.

In an effort to separate well depth from intensity of use, an ordinary least squares regression was done, with wells per rig year as the dependent variable. The Department of Energy (DOE) revised well completion series was used for 1973–1985, and estimated back to 1967 by applying the 1973–1977 ratio of the DOE series to the API series (1.055). The independent variables are (1) the ratio of active rigs to all rigs, as tabulated starting 1967 by Reed Tool Co., and (2) the average depth of well. If WRY = Wells per Rig Year, PAR = Percent of Active Rigs, and AWD = Average Well Depth, then the estimating equation is:

$$WRY = 14.3 - 0.49(\ln PAR) - 1.32(\ln AWD) \qquad (5)$$

Thus for every additional percent of rig capacity utilized, the number of wells per rig year fell by 0.49 percent; for every additional percent of average well depth, wells per rig year decreased 1.32 percent. The R^2 was .78, F-statistic 32, respective t-statistics 3.8, 4.9, 3.0. This looks robust, but the extremely low Durbin–Watson (0.95) downgrades the significance. As usual, the small sample and much collinearity take their toll. However, it seems clear that there has been a great short-term gain in efficiency, which may have more than undone the waste and misdirected effort of the 1970s.

One hesitates to credit a doubling of efficiency (wells or footage drilled per rig year) in only the five years, 1981–1986. The mix of wells, as between oil and gas, deep and shallow, and so forth must have changed. But a special tabulation done at Baker Hughes, which allows for these changes, shows an increase in efficiency of over 70 percent. My own conjecture is that these gains were made slowly and incrementally since 1973, but were masked by the gross inefficiencies unleashed by the drilling boom. As activity slumped after 1981, the gains quickly appeared. We should not, therefore, expect to see them continue for long.

1. An Attempted Explanation

We now try to draw the threads together, and first try to explain to some degree why costs were stable or possibly even decreasing during 1955–1972. It takes some explaining because cumulative output went from 33.2 billion barrels end–1946 to 103.4 end–1973. Proved reserves of 21.5 billion at the end of 1946 were used up and replaced three times over during the next 27 years, each replacement from fields inferior to the previous. An exogenous factor was the decline, after 1949, in the real price of oil in the United States, following trends in the world price, although it was always considerably higher.

Imports were subject to quotas, first informal, then "voluntary," then mandatory. But I think it was well understood by the late 1950s that further price increases would not be tolerated, and might lead to loss of quota protection. Indeed, the struggle over oil imports was never-ending. In 1969, the Nixon administration undertook a review of the whole import question, by a task force under the general direction of George P. Shultz, which favored abandonment (Cabinet Task Force on Oil Import Control, 1970).

The import quota had sheltered a system of market demand prorationing which restricted output, and favored small unproductive wells. It was often profitable to drill many more wells than needed to drain a reservoir optimally. The additional wells brought no additional capacity, only increased production "allowables" (Adelman, 1964).[6]

As the real price slowly declined, these wells were curtailed. Moreover, poorer prospects were no longer drilled. As Column 5 in Table 3 shows, the number of oil development wells fell drastically, from 29 thousand in 1955 to 9 thousand in 1973. (Gas development wells actually rose.) By that time, production was no longer restricted by market-demand prorationing.

The result was a one-time gain in efficiency, the precise amount of which is impossible to measure. This must account for some part of the substantial oil cost reduction during a period of massive resource depletion.

However, from 1973 through 1976, the oil industry reacted to the increase in oil prices by bidding up factor prices even faster; while real costs rose faster yet. It is a tribute to the power of expectations, and of course a classic example of the accelerator–multiplier principle.

By 1978, this imbalance had been largely corrected, and the price-cost margin was again at about the 1973 level. But then came the even larger price jump of 1979–1981 and an even stronger industry reaction. Nothing was too extravagant, since everyone knew that the price of oil was going to $100 or $200 by the end of the century. At the same time, gas price regulation had generated excess demand, and focused all of it upon a small number of exempt sources of supply: recently or newly found gas, imports, and deep gas whose price touched $9 per mcf.

The bubbles began to burst after 1981, but the effect was two sided. Footage drilled per rig year, for example, which had fallen from 125 thousand in 1972 to 102 thousand in 1981, rose to over 200 thousand at the end of 1986, and was still rising in early 1987. To be sure, such comparisons are always distorted by the shift in the mix of wells, oil and gas, shallow and deep, productive, and dry.

In any case, the fall in factor prices was such that the producer's real price of oil actually increased, while development cost fell again. By 1984, it was at about the 1955–1963 level. Despite the sharp decline in rigs running, there was only a mild decrease in oil development wells drilled. The volume of oil reserves-added in 1983–1985 was a near-record for any three years, exceeded in the lower 48 only in 1949–1951[7] (API, 1959, 1979; EIA, 1986). Cost inflation and inefficiency cut so sharply into the incentive of higher prices, that it is hard to discern the net effect of the price increase. Conversely, the price decrease since 1981 has been buffered and offset by the decrease in costs.

Multiple regression analysis is the standard way of sorting out such effects, but the barriers are formidable. The obvious candidates for independent variables are the number of oil development wells and/or some other index of intensity used to register the movement along the supply curve; cumulative production to register the depletion effect which displaces the whole supply curve; and the depletion rate or production/reserve ratio pertains to both kinds of effects. The higher the depletion rate, the greater the shift away from exploration and toward development as a source of reserves. It was pointed out earlier that these were substitutes, hence proxies, because the more expensive it became to find new fields, the greater the inducement to increase drilling in and around old fields, and produce at higher depletion rates.

But all these variables are highly intercorrelated, and each is serially correlated with itself. For the period before 1974, no regression showed any relation. The appearance of a relation comes only when one includes the later years, which makes the results unacceptable for any attempt to explain the depletion effect upon costs. Possibly a much larger scale study, combining time series with cross-sections by states, could fill the gap.

2. Regression Study of Reserves Added

Table 5 shows the results of a somewhat different approach: explaining the changes in annual crude oil reserves-added by cost and price factors.

Table 5. Regression Results: Determinants of Reserves Added

No.	Constant	Cumulative Output	(Cumulative Output)^2	Drilling Factor Prices	Nominal Price	Adj. RSQ	F-stat	D-W

Logarithmic Equations

1	3.69	—	—	−0.198	—	0.156	8.24	1.14
	(34.04)	—	—	(2.87)	—			
2	3.45	—	—	—	−0.092	0.042	2.7	1.04
	(81.00)	—	—	—	(1.64)			
3	5.05	−0.34	—	—	—	0.197	10.54	1.22
	(9.87)	(3.25)	—	—	—			
4	6.00	−0.54	4.590E−22	—	—	0.205	6.02	1.27
	(6.32)	(2.71)	(1.18)	—	—			
5	3.42	0.37	—	−1.58	0.928	0.481	13.02	1.62
	(4.09)	(1.54)	—	(4.5)	(4.5)			
6	3.52	0.04	1.540E−22	−1.57	0.9	0.466	9.52	1.62
	(3.65)	(1.37)	(0.21)	(4.44)	(3.76)			
7	4.67	—	—	−1.14	0.75	0.461	17.7	1.57
	(21.08)	—	—	(5.53)	(4.47)			

Arithmetic Equations

8	2960	—	—	−6.31	—	0.068	3.8	1.23
	(15.8)	—	—	(1.96)	—			
9	2621	—	—	—	−6.07	0	0.19	1.15
	(15.8)	—	—	—	(0.45)			
10	3377	−0.01	—	—	—	0.164	8.6	4.41
	(11.6)	(2.94)	—	—	—			
11	4734	−0.048	2.30E−07	—	—	2.12	6.24	1.54
	(5.94)	(2.26)	(1.82)	—	—			
12	3617	−0.0065	—	−34.3	141.8	0.459	12	1.98
	(12.9)	(1.18)	—	(3.52)	(4.65)			
13	3790	−0.015	3.60E−08	−34.7	139.1	0.445	8.81	1.98
	(4.34)	(0.47)	(0.21)	(3.45)	(4.01)			
14	3351	—	—	−42.1	153.3	0.453	17.2	1.88
	(20.1)	—	—	(5.83)	(5.27)			

Note: *t*-statistic in parentheses.

Equations (5)–(7) and (12)–(14) show the explanatory variables as the nominal price of crude oil, and the IPAA drilling cost (factor price). In both the arithmetic and logarithmic version, the factor price variable is stronger. Supposing that both variables double, "real price," as defined here, would be unchanged. However, because $2^{-1.23} * 2^{-.74} = 0.71$, reserves-added would drop by 29 percent. Apparently, the drilling cost variable incorporates not only the effect of drilling cost, but also the effect of lower efficiency, with which it is strongly correlated.

We must recognize an identification problem. Reserve-increments (the dependent variable) have no effect on the price of oil, nor on cumulative depletion. But they do have an effect on the index of factor prices (IPAA). The greater the inputs into oil development, the greater the reserve increments, as well as the greater the pressure on factor prices. This may be tolerable because most inputs are on exploration and on gas development. But to some extent it impairs the relationships measured in Table 5.

Equations (3)–(6) and (10)–(13) incorporate the effect of cumulative production. Taken in isolation, Equations (3) and (10) have the right sign and a good t-statistic. But when allowance is made for the effect of prices and costs, the coefficients are small and insignificant. The effect is complex. Cumulative depletion registers cumulative learning. Hence the attempt to use a quadratic function to capture also the ultimate resource–exhaustion effect. But nothing can be observed.

We look now at the outstanding anomaly, the extraordinarily bad performance of the mid-1970s, when a huge upsurge of exploratory and development drilling coincided with a very low level of reserves-added. There would be no mystery if the drilling boom were exogenous, and simply imposed higher money and real costs. But this is not so; the drilling upsurge was a response to perceived profit opportunities. It is difficult to get inside the mystery because of the wide variation in oil prices at any given moment. For many leases whose output was under price control further development drilling was unprofitable. Yet for the whole producing system, the desired level of drilling plainly exceeded what could be done.

Analysis is the more difficult because after 1973, averages of price and cost become unreliable because of price regulation. Moreover, the reserves-added data are also less reliable.

Figures 8 and 9 show another physical indicator of cost: average daily output per oil well and per gas well. In oil, there is a slow increase after 1958, peaking in 1973 and declining thereafter, but still well above anything before 1967. In gas, there was a persistent decline in California and Texas. The increase in Louisiana reflects the growing importance of offshore production. The Louisiana decline after 1974 shows a substantial decline in efficiency, or increase in cost.

E. Finding Costs, Development Costs, and Resource Values: 1970–1985

The value of any asset, in any industry, is equal to the *lesser* of (a) the present discounted value of its future surplus over operating costs, or (b)

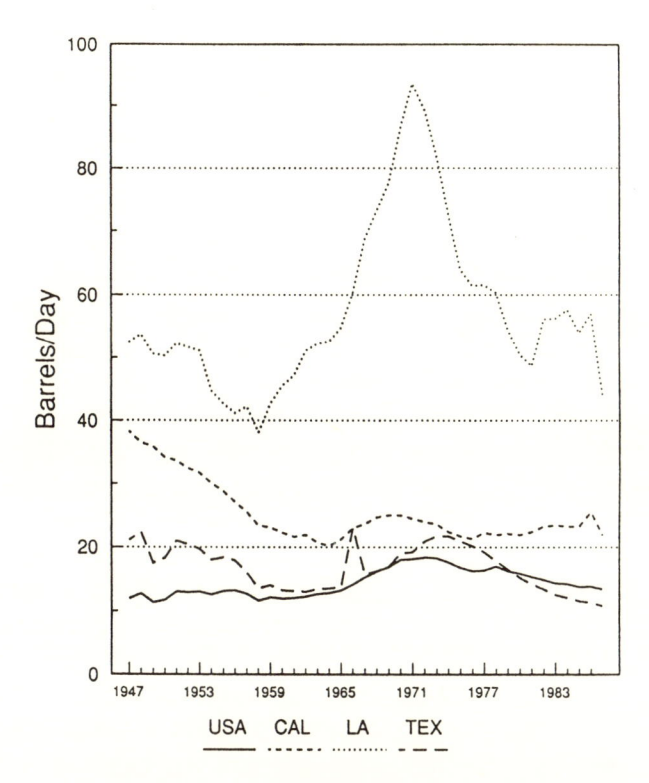

Figure 8. Average Oil Production per Well (Barrels/Day)

its replacement cost. This ratio has become well known as "Tobin's Q." In a mining industry, as we have emphasized, replacement may be had by (b_1) development or (b_2) exploration.

A developed or undeveloped barrel in the ground is an asset, governed by the general rule. Value and replacement cost are always gravitating toward each other. The higher the value, the greater the finding incentive and investment. This drives up the cost. Conversely, the higher the cost of finding, the greater the value of a barrel already found.

In whatever industry, nothing is more difficult than predicting future returns from holding an asset for later sale. Yet owners must make the comparison with replacement cost as best they can, and decide whether or not to invest or to is disinvest, that is, sell off some of the asset.

When oil prices are under pressure, we hear ad nauseam about the foolish oilmen who will sell off "the incremental barrel" at anything over

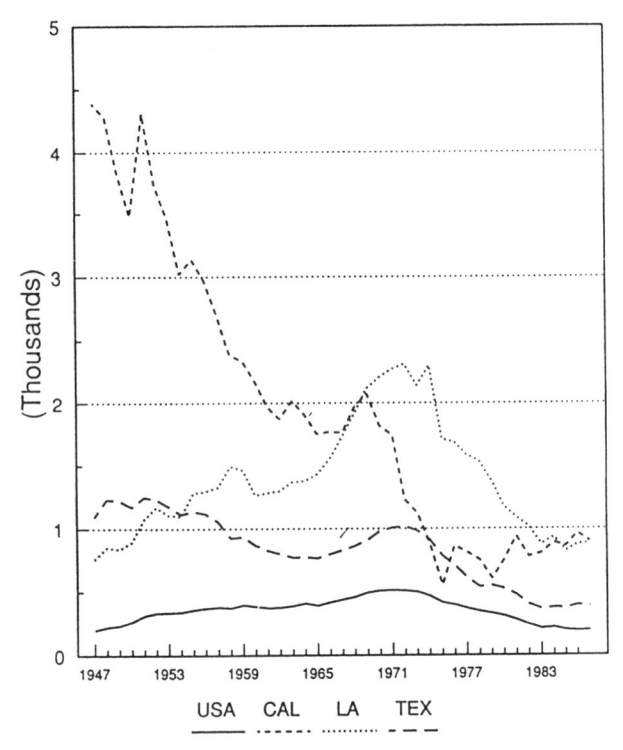

Figure 9. Average Gas Production per Well (MCF/Day)

bare operating cost. In fact, they will do no such thing, unless they think the whole industry is liquidating, and prices will never recover. Otherwise they must take into account the value of the asset they consume: reserves.

Prices soften or decline when they are too far above the *total* of operating costs *plus* development cost *plus* the lesser of resource value or marginal finding cost. We should not lose sight of the fact that in the areas holding most of the world's reserves, a price as low as $10 is still very high compared with the sum of the three. Our current concern, however, is with the United States.

We propose to show the relation by first estimating what price above ground would barely compensate the investment calculated here, and comparing this supply price with actual above-ground prices. Table 6 shows oil and gas revenues during 1955–1983, first gross then net of operating costs, taxes, and royalties. The ratio stayed within a narrow

Table 6. Production Expenses, Taxes and Royalties, Oil and Gas: United States, 1955–1986
(dollars in millions)

Year	Gross Revenues	Net Revenues	Ratio Net to Gross
1955	7848	4895	0.624
1956	8347	5153	0.617
1967	N.A.	N.A.	N.A.
1958	N.A.	N.A.	N.A.
1959	9031	5705	0.632
1960	9211	5964	0.648
1961	9562	6187	0.647
1962	9919	6401	0.645
1963	10295	6673	0.648
1964	10405	6731	0.647
1965	10653	6815	0.640
1966	11429	7152	0.626
1967	12274	7795	0.635
1968	12964	8136	0.628
1969	13882	8755	0.631
1970	14919	9353	0.627
1971	15789	9896	0.627
1972	15893	9918	0.624
1973	17952	11383	0.634
1974	28154	18724	0.665
1975	32061	21433	0.669
1975	32798	21908	0.668
1979	36709	24418	0.665
1977	42445	28346	0.668
1978	47850	31718	0.663
1979	60538	40819	0.674
1980	93629	59048	0.631
1981	135615	74711	0.551
1982	128195	73220	0.571
1982	135590	79896	0.589
1983	126620	72173	0.570
1984	132590	75576	0.570
1985	122050	69569	0.570
1986	75060	48414	0.645
1955–1980 Average			0.645
S.D.			0.016
1981–1982 Average			0.570
S.D.			0.016

Sources:

Gross Revenues:

1955–1975: Joint Association Survey, Part II

1975*, 1976–1986: Bureau of the Census, Annual Survey of Oil and Gas.

1982*, 1983–1986: Department of Energy, *Annual Energy Review.*

Net Revenues:

1955–1982: Gross Revenues minus royalties, production outlays, and taxes, from J.A.S., 1955–1975; from Census, 1975*–1982.
Net Royalties assumed 15 percent of gross revenues, less royalties recieved by operators. Net royalty averaged 12.7 percent 1955–1982, assumed so for 1982*.

1983–1985: Ratio of Net to Gross for 1981–1982 used to account for Windfall Profits Tax in effect.

1986: Ratio for 1955–1980 used, assuming the Windfall Profits Tax not in effect due to fall in prices.

range of 35 to 37 percent before 1974, but then declined mildly as prices outran costs, and then jumped with the imposition of the misnamed Windfall Profits Tax in 1981.

We turn now to the comparison of in-ground values with prices and costs. Published estimates of current values are based on current sales of properties. The sample for any one year is very small. The dispersion is great, since each sale is a single payment for a bundle of oil, gas, and many special features, good or bad. It becomes hard to discern the values of the oil or gas reserves as such. Moreover, published results are "oil equivalent barrels," mingling oil and gas together.

A recent paper (Adelman, DeSilva, and Koehn, 1989) tabulates the individual company estimates issued by the John S. Herold Company beginning 1946. They calculate the present value of the proved reserves, expected to be depleted in some trajectory over time, and discounted at what is considered an appropriate rate.

The Herold valuations are not market data. However, they are themselves subject to a market process: the nearer they come to what investors consider reasonable—that is, would pay or demand—the more successful they are. The survival of the company for this long a period indicates that they have been acceptably close; and the Herold valuations are frequently referred to and quoted in the financial press.

In Table 7, columns 1 and 2 show the average wellhead price of crude oil, and the price net of operating expenses, royalties, State severance taxes, and (in recent years) the Windfall Profits Tax, which is an excise not a profits tax. For 1955–1982, there are actual data on these deductions. Since as shown in the previous table the dispersion about the period mean is quite small, it seems safe to extend it forward and back.

Column 3 shows the annual average value of proved reserves of crude oil in the United States. Outside this country, not only are reserves calculated differently, but the risk and discounting factors are different.

Figure 10 shows the salient values: prices, values and costs over a 40–year period. These figures are all in nominal (current) dollars. In real (inflation-adjusted) terms, they all decreased through 1973.

For many years, the industry has had a rule of thumb of in-ground value as one-third of the wellhead price. It seems to be well supported. Figure 11 (lower line) shows value is mildly below one-third before 1973, mildly higher afterward.

Column 4 of Table 7 recapitulates the development cost estimates from earlier tables. Column 5 adjusts it for tax benefits which lowered

Table 7. Wellhead Price, Cost, and Reseve Values: United States, 1946–1986 (Current Dollars per Barrel)

Year	Wellhead Price Gross	Wellhead Price Net	Average Reserve Value	Development Cost Pre-tax	Development Cost Post-tax	User cost (Value less post-tax cost)
1946	1.41	0.91	0.30			
1947	1.93	1.24	0.43			
1948	2.60	1.67	0.75			
1949	2.54	1.64	0.73			
1950	2.51	1.62	0.70			
1951	2.53	1.63	0.73			
1952	2.53	1.63	0.68			
1953	2.68	1.73	0.75			
1954	2.78	1.79	0.69			
1955	2.77	1.78	0.89	0.80	0.71	0.18
1956	2.79	1.80	0.95	0.80	0.72	0.23
1957	3.09	1.99	0.94	n.a.	n.a.	n.a.
1958	3.01	1.94	0.91	n.a.	n.a.	n.a.
1959	2.90	1.87	0.88	0.50	0.45	0.44
1960	2.88	1.85	0.83	0.66	0.58	0.24
1961	2.89	1.86	0.83	0.58	0.52	0.31
1962	2.90	1.87	0.87	0.77	0.69	0.18
1963	2.89	1.86	0.83	0.74	0.66	0.17
1964	2.88	1.85	0.87	0.61	0.55	0.32
1965	2.86	1.84	0.83	0.53	0.47	0.36
1966	2.88	1.85	0.84	0.54	0.48	0.36
1967	2.92	1.88	0.82	0.56	0.50	0.32
1968	2.94	1.89	0.83	0.71	0.63	0.20
1969	3.09	1.99	0.85	0.85	0.75	0.10
1970	3.18	2.05	0.81	0.63	0.56	0.25
1971	3.39	2.18	0.90	0.68	0.61	0.29
1972	3.39	2.18	0.99	1.10	0.98	0.01
1973	3.89	2.51	1.41	0.90	0.80	0.61
1974	6.74	4.34	2.28	1.30	1.16	1.12
1975	7.56	4.87	2.39	3.09	2.75	–0.37
1976	8.19	5.27	2.74	4.17	3.71	–0.97
1977	8.57	5.52	3.03	3.83	3.41	–0.38
1978	9.00	5.80	3.39	2.75	2.45	0.94
1979	12.64	8.14	4.62	3.96	3.52	1.10
1980	21.59	13.90	7.91	3.48	3.10	4.81
1981	31.77	17.51	9.72	6.42	5.71	4.01
1982	28.52	16.28	8.31	11.29	10.05	–1.74
1983	26.19	15.43	8.52	4.50	4.01	4.51
1984	25.88	16.67	8.80	3.63	3.23	5.57
1985	24.09	13.80	8.19	4.08	3.63	4.56
1986	12.51	8.15	5.88	4.96	4.42	1.46

Sources:

Gross price: Table 4.

Net price: Gross price times ratio of net to gross revenues from Table 6.

Average reserve value: Adelman, DeSilva, and Koehn (1987).

Development cost:
  Pre-tax: Calculated from Unit cost in constant dollars given in Table 2 using the Drilling cost index from Table 1.
  Post-tax: See below.

User cost: Average reserve value minus Post-tax development cost.

Post-tax costs: The reduction in cost aims to capture the net advantage of drilling for oil instead of buying. This is the result of the tax advantage of charging off intangible drilling expenses. The percentage depletion allowance, repealed for nearly all properties in 1975, affected the value of a property, whether obtained by buying or by drilling. Intangible drilling costs are "between 60 and 70 percent of the entire well cost" (*Petroleum Production Handbook*, 1962, pp. 38–22; repeated at pp. 44–11 of *Petroleum Engineering, Handbook*, 1987). However, a special API tabulation released in 1985 showed intangibles as 34 percent in 1984. This discrepancy is due to the fact that drilling and completion account for only about 60 percent of total development cost including lease equipment, pressure maintenance programs, etc. For the whole period, therefore, development outlay post tax is reckoned at 83 percent of pre-tax, by the formula: $X = 1$ minus $(.34)(1$ minus $.5) = .83$. The net present value is 63 percent of the gross saving. This is calculated by assuming that cost would otherwise be uniformly charged off over 25 years, and discounting at 10 percent, this would be worth 0.367 of an immediate payment, that is, 1 minus .367 = .633. Then 0.17(.63) = 0.11, and development cost is reduced by 11 percent.

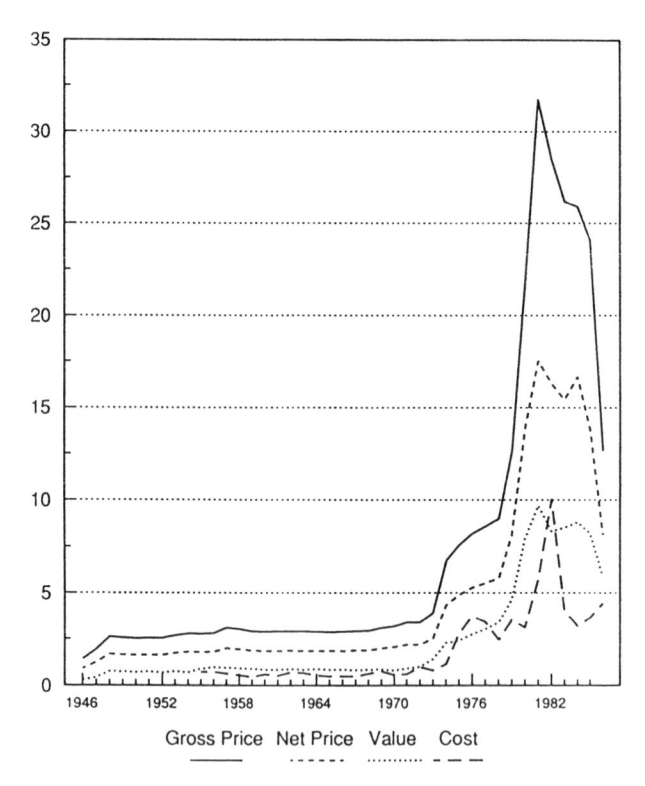

Figure 10. Crude Oil: Price, In-Ground Value, Cost (United States, 1946–1986)

the net cost to the investor, more substantially before 1975, when percentage depletion was repealed. Since value is also after tax, this is necessary to make the cost comparable with the value. The average ratio of value to post-tax cost over the whole period 1955–1986 is 1.6. (The only years with values below unity are, significantly, during the very disturbed if profitable years since the first oil shock.)

The excess over unity, stated in column 6, is equal to resource rent or user cost, the present value of future use. It is the pure resource value sacrificed by choosing to develop today rather than later. A developed reserve is worth the present value of revenues less operating outlays, but an undeveloped reserve is worth only revenues less the sum of operating outlays plus development cost. Therefore, if one subtracts out development cost per unit from reserve value, one has the pure resource value of a unit in the ground. In equilibrium, resource rent equals discovery cost,

Finding and Developing Costs in the United States 45

Figure 11. Crude Oil: Value in Ground as Percent of Price
(United States, 1946–1986)

and should indeed be regarded as a rough estimate, especially when averaged over a period of years.

In nominal terms resource rent was quite stable through 1973, and paralleled development cost. This confirms what we concluded in reviewing the theory: development cost and exploration cost move in parallel because they are substitutes for each other. The years 1975, 1976, 1977, and 1982, showed negative user cost. These were all unusually bad years in terms of reserves-added.

Figure 13 shows "Tobin's Q": the relation of value and development unit cost. In theory, the value of a developed barrel is simply the discounted present value of the oil (net of current operating expenses and taxes) which is to be produced out of existing installations. The value of an undeveloped barrel should equal the present value of the oil less the development investment. This is its resource rent or user cost.

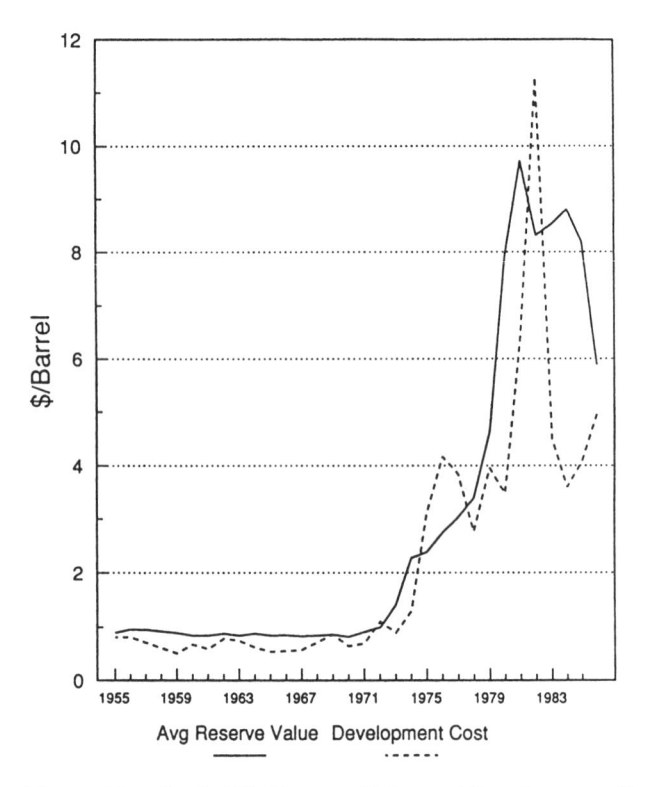

Figure 12. Crude Oil: Reserve Value and Development Cost
(United States, 1955–1986)

For any given oil lease, or for all of them taken together when they are all in equilibrium, the value of an additional undeveloped barrel cannot exceed its nearest alternative: the avoided cost of developing the existing reservoir more intensively. An operator will not pay more for a new undeveloped barrel, whether by discovery or purchase, than he can obtain one by development of an existing reservoir. But this in turn depends on the intensity of its current development.

At one extreme, there is zero resource constraint and additional oil can be developed at the same cost as the current increment. Here it does not pay to discover or acquire an additional reservoir, because there is no penalty in working existing reservoirs more intensively. At the other extreme, existing reserves cannot be expanded at all. Additional output can only be had by more intensive development.

A simple model of reservoir development costs is given in the Appen-

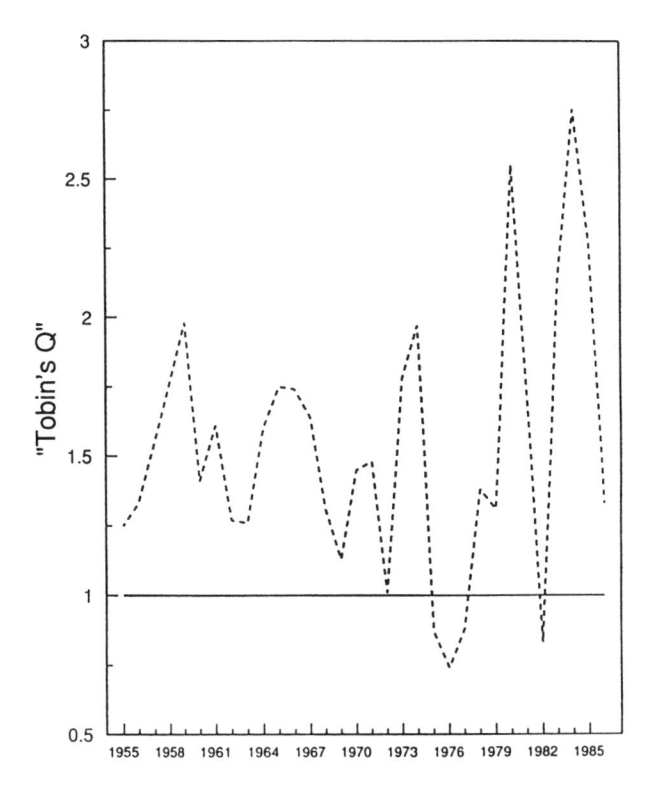

Finding and Developing Costs in the United States 47

Figure 13. Crude Oil: Value Relative to Development Cost
(United States, 1955–1986)

dix (for full treatment, see Adelman, 1990). Since we take marginal cost to be a linear function of intensity, the addition to marginal development cost lies between zero and unity. Hence the value of an undeveloped barrel will also stay within these limits. The value of a developed barrel should then stay between 1.0 and 2.0 times the level of development unit cost.

Figure 13 shows that the average value of a barrel did stay within those limits through 1973. The mean ratio was 1.47. During 1974–1986, the mean ratio was higher, at 1.62, but it was much more variable, staying between 1.0 and 2.0 only five years out of thirteen.

1. Price Expectations and Asset Values

In general, a long-lived asset rises in price when the market expects an increase in the prices of the goods in which the asset will be embedded through future production. Hence asset price changes are a leading indicator or forecaster of product price changes. There was no indication of higher user cost, hence future higher prices, before 1973, which suggests that there was no room for higher prices within the framework of expected supply and demand.

Table 7 permits us to restate Equation (4):

$$V = Pa/(a + i - g) \tag{6}$$

where the net present value of a barrel in-ground is the net price above-ground, discounted for the cost of holding the undepleted portion over the remaining lifetime of the pool. In Equation (6), we have introduced a new parameter g, the rate at which the price is expected to rise.

According to the well-known "Hotelling valuation principle" (HVP) (Miller and Upton, 1985), the price of a mineral is expected to rise at the riskless interest rate, since at a lower rate the owner of the deposit would be losing what he could earn by selling the mineral and investing the proceeds. But a given percent rise in the price is equivalent to a decline in the interest rate. Hence in equilibrium $g = i$. If so, Equation (6) reduces to $V = P$, the value of the deposit equals the current undiscounted price. The elegant simplicity of this principle is attractive. But it is contradicted by the data of Table 7 (see also Figure 11).

The HVP cannot work because the owner of the reserve cannot in fact extract and sell it off entire. At most he can deplete a minor fraction in any given year. The faster he depletes, the higher the present value. This relation is embodied in the reservoir engineer's "deferment factor," which is proportional to the reserve-production ratio. The lower the deferment factor, the higher the ratio of discounted to undiscounted value (*Production Engineering Handbook*, 1987). But the lower the deferment ratio, the higher the cost. Optimal depletion, therefore, involves a tradeoff. We can see how the one-third rule embodies this fact if in Equation (6) we set g to zero, and assume that the decline rate and the discount are approximately the same. In that case, $V = P/2$.

However, that is only a broad generalization approximately true over a long period. As noted above, in-ground value is an index of price expectations. If the net prices are expected to rise in the future, the present value of a known deposit rises immediately. Therefore an increase in the ratio of V to P signals an expected rise in price. Rearranging Equation (6) we have:

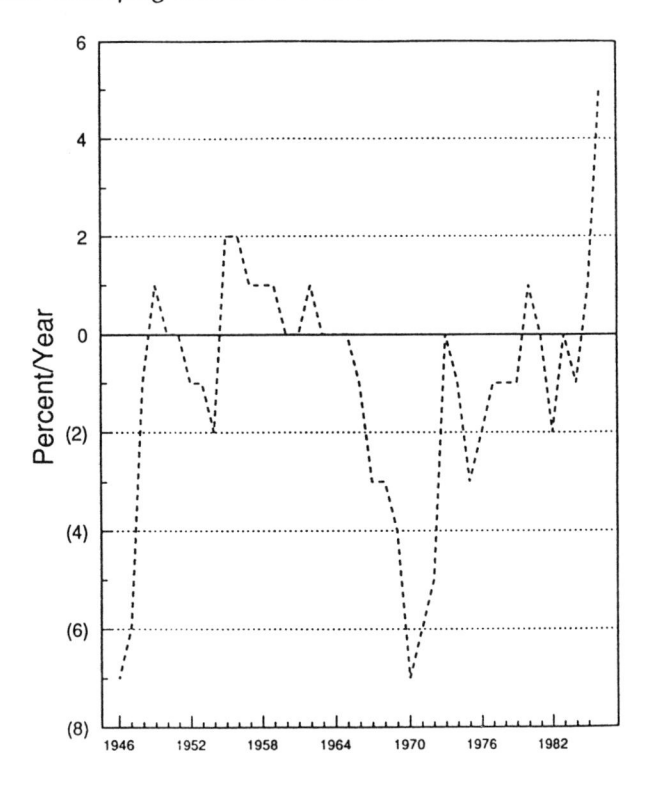

Finding and Developing Costs in the United States

Figure 14. Expected Real Price Increase
(Discounting @ 10%)

$$g = i + a(1 - (P/V)) \qquad (7)$$

We can make this calculation from the data in Table 7, and the annual production:reserve ratio in Table 3, by assuming the real discount rate is 10 percent. It is shown in Figure 14.

We cannot claim accuracy for these estimates, since they are very sensitive to the discount rate. However, it is easy to substitute a better one; there is a one-to-one correspondence between changes in i and changes in g. It would raise or lower the curve, but not change its slope anywhere.

Figure 14 shows that the steep price increases of 1946 and 1947 were not expected to last; they did last, however. There was some mild pessimism in the early 1950s, which was reversed in 1955; the better tone lasted through the early 60s, but then became increasingly poor. Ever-worse price deterioration was expected through 1970. Here the apprehension was justified, since the real price kept dropping. But expectations grew less bad in 1971–1972, and then became practically neutral in 1973. The poor expectations of 1974–1979 indicated the industry did not expect to keep all of its gains, but it is interesting that when the price shot up in 1980–1981 expectations were for further increase. But the turnaround in prices did not lead to bad price expectations. Interestingly, the price deterioration of 1985 improved g, and the 1986 collapse put it to an unprecedentedly high level. In other words, the low 1986 price was expected to turn around, and this was a correct forecast.

V. CONCLUSIONS

Since the whole earth is finite, any mineral in it is finite, but we know not where the limits are, and it would not matter if we did. We will never get to the end of our oil resources. We will stop impounding them into reserves when it no longer pays.

To treat the total of "economic resources," that is, those worth producing, as a prefixed nonrenewable stock, is circular reasoning. It assumes the conclusion. For in order to estimate "economic resources" we first need to estimate future costs and prices. We might as well claim that there was a prefixed number of buggy whips to produce. In fact, the logistical curve is a good description of a manufacturing industry, every one of which has gone through a phase of accelerating then decelerating growth (Burns, 1934).

Cost increases would be a sonar "ping" warning us we were getting closer to the end, when it would no longer pay to find and develop. The ping did not get any louder during 1945–1973. Development cost tended if anything to decrease, an indicator that discovery costs were not rising either. It roughly doubled after 1973 returning to the 1955–1962 level. Much but by no means all of the increase is explained by the inefficiencies and waste imposed by a more than threefold investment expansion in only eight years.

We are left with a realization of how much reserves can keep expanding literally decades after all the big fields are found, and at no increase in real cost. That expansibility is now being sorely tried.

Meyer, Woods, and others have drawn attention to the estimates derived from discovery process models of very large numbers of very small fields, relatively easily found, and containing, in the aggregate, large amounts of oil (Meyer and Fleming, 1985; Woods, 1985; GRI,

1985). Smith and Paddock (1984) have been able to approximate discovery decline curves in various provinces. A small difference in the slope of the decline curve of field size makes a considerable difference in the total area to be added. And both the height and slope of the discovery curve depends on improved development technology, which improves finding rates, because more new fields are worth exploitation.

Cumulative depletion has its favorable side. Intensive drilling and production have produced a dense network of pipelines and infrastructure, which allows quick production, thereby lowering cost. One sees a similar development in places like the North Sea, where pipelines to shore were a major fraction of development cost, and remain to be used by newer smaller fields as the older larger ones go into decline. Moreover, development costs per drilling-production unit in 1,000 feet of water have dropped about three-fourths in ten years (Petrie and Wright, 1985).

In Alaska, Prudhoe Bay had about 25 billion barrels in place, of which 10 billion became reserves. But overlaying Prudhoe Bay are the West Sak and Ugnu formations containing perhaps 70 billion barrels in place (OGJ, 1985b). If any of it is developed into reserves—none has been—it will certainly be much higher-cost oil than Prudhoe. On the other hand, the Kuparuk waterflood promises one billion barrels for $445 million, cheap in any league (PIW, 1985).

In mid-1987, a developed barrel of oil reserve in the ground appeared to be worth about $4.50, while its cost was about $3.00. On average, therefore, the industry is in no danger of disappearing, nor even of drastic shrinkage. But a considerable number of marginal deposits have lower value and higher cost, and will be scratched. Moreover, if the pure resource value is $1.50, one must question how much oil can be discovered at that cost or less. Without new fields and pools to freshen the mix, development costs must creep upward; however, we cannot tell how fast.

The collapse and recovery of oil prices in 1986–1987 begins a new chapter. For the first quarter of 1987, the number of rotary rigs operating averaged 825, about half of the 1975 level. But the number of wells completed, and footage, were about equal to 1975, while prices in the $15–$20 range compared with $5–$12, average $7.67 (Department of Energy, 1975).

Today, the better prospects look better than ever, at the expense of a lot of poor or mediocre ones. It is reasonable to expect a shrinking domestic industry at prices in the $15–$20 range, but we will hazard the guess that the decline will be quite slow. The forecasts of the National Petroleum Council and others (PIW, October 1986), that production in and outside the United States will shrink even as prices rise, may turn out to be true, but as of now they have no foundation in fact.

APPENDIX

Cost Calculations

The objective of the calculations is to obtain the marginal or incremental capital cost per barrel of oil or per mcf of gas by investing in a new project. Define:

P = market price *or* supply price (see below)
i = market interest rate
K = capital expenditures
R = reserves to be developed
Q = initial or peak output
a = exponential decline rate
T = life of project

Taking i the discount rate as exogenous, we calculate P as supply price or cost, the price at which the investment would be just barely worth making.

Derivations

(1) Reserves are cumulative production, declining exponentially:

$$R = Q \int_{T}^{t=0} e^{-at}\, dt = Q\, \frac{1 - e^{-at}}{a} \tag{A1}$$

For most values of a and T, $(1 - e^{-aT})$ converges quickly to unity, and may usually but not always be dropped.

(2) For a given investment, the Net Present Value:

$$(\text{NPV}) = PQ \int_{t=0}^{T} e^{-(a+i)t}\, dt - K = 0$$

Finding and Developing Costs in the United States

$$PQ(1 - e^{-(a+i)T}) = K(a+i) \tag{A2}$$

Substituting Equation (A1) into this equation, and transposing:

$$P = (K/R)\frac{a+1}{a}\frac{(1-e^{-aT})}{(1-e^{-(a+i)T})} \tag{A3}$$

For typical values of a, i, and T, the last fraction converges quickly to unity, and may usually but not always be dropped, *or*

$$P = (K/(Q/a))(a+i)/a = (K/Q)(a+i) \tag{A3a}$$

In words, the supply price is equal to (a) the cost per unit in-ground (K/R) multiplied by the adjustment for holding the stock until produced, $(a+i)/a = (1 + (i/a))$; *or* to (b) the investment per annual barrel multiplied by the compound discount rate.

Alternatively, if the price is known, we solve for the rate of return:

$$i = a((PR/K)-1) \tag{A4}$$

or,

$$i = PQ/K - a \tag{A4a}$$

Where T is known, find a by solving Equation (A1), but using actual T instead of infinity. Then insert a and known T into Equations (A3) or (A4), and solve alternatively for P or i.

(3) In the foregoing, we assume all capital expenditures made at one time, in year zero. We assume peak output initially, then an exponential decline.

In fact, capital expenditures stretch over several years, usually peaking in or just before year zero, when production starts. Typically, production builds up over 2–3 years, then holds approximately stable for a few more, then declines steeply.

The errors are mutually offsetting (Adelman and Paddock 1980) showed that for North Sea fields, the value of P calculated as above gave an excellent but somewhat down-biased prediction of the values as calculated from the actual production plans, tabulated by Wood McKenzie.

The result of the Adelman-Paddock test is not surprising. If T is finite, then from Equation (A1) the final-year output $Q_f = Q_0 e^{-aT}$, and $a = (Q_0 - Q_f)/R$. Then it follows that:

$$T = \frac{R}{Q_0 - Q_f} \ln(Q_0/Q_f) \qquad (A5)$$

where T is the period in years, R = reserves = estimated cumulative output, Q_0 = initial-year output, and Q_f = final-year output.

If T is finite, a is less, and the extreme right-hand fraction of Equation (A4) is less than unity. Hence the supply price is less, and the return is higher, than would result from our simpler use of the (K/R) $(a + i)/a$ approximation.

Consider proved reserves of 100, initial-year output of 10. On the usual assumptions of infinite life, the depletion/decline rate is taken at $10/100 = 0.1$. Assume $(K/R) = \$1$, and the discount rate at 10 percent. Then by Equation (A4), with the final right-hand fraction converging to unity:

$$P = \frac{K}{R} \frac{a+i}{a} = 1*2 = \$2 \qquad i = [Pa/(K/R)] - a \qquad (A6)$$

Alternatively, assume that the market price $P = \$3$, $i = 0.2$. Suppose we know, however, that final-year output is 2. Then we calculate

$$T = 100/(10 - 2) (\ln 5) = (12.5)(1.6) = 20.1 \text{ years}$$

If so, a decline rate of 8.0 percent per year yields cumulative output of 99.96, and $P = \$1.85$. Alternatively, if we assume a market price of \$3, then on the infinite-life assumption, the rate of return is 20 percent. Setting $T = 20.1$ and $a = .08$, the resulting $i = 22$ percent.

Thus, our method, which assumes infinite time, understates present value, and overstates cost by a factor of 9.2 percent.

(4) We assume a developed reservoir, operating or about to start. As previously,

$$Q = \text{initial output}, R = \text{reserves} = Q \int_{t=0}^{T} e^{-aT} dt$$

We add the definition

$$K = \text{capital expenditures} = kaQ, \text{ or } K/Q = ka$$

where k is an empirical constant reflecting better or poorer geology.

For an additional tranche, make two alternative assumptions: (1) reserves expand at the same rate as output; *or* (2) reserves do not expand at all. (We do not consider the case of reserves expanding faster than output.)

Alternative 1. There is no resource limitation, and the long run marginal cost curve is horizontal. Accordingly:

$$dQ/Q = dR/R, \text{ hence } dR/dQ = R/G$$

Since $a = Q/R$

$$da/dQ = \frac{R - Q(dR/dQ)}{R^2} = \frac{R - Q(R/Q)}{R^2} = 0$$

$$dK/dQ = d(kaQ)/dQ = ka + kQ(da/dQ) = ka \quad \text{(A7a)}$$

Therefore investment per additional annual barrel is unchanged, and the value of an undeveloped barrel is zero.

Alternative 2. Reserves are fixed. Additional investment can only accelerate output, not increase the total over time. Then the factor a must increase proportionately with Q:

$$dR/dQ = 0$$

$$da/dQ = \frac{R - Q(dR/dQ)}{R^2} = 1/R = a/Q$$

$$dK/dQ = d(kaQ)/dQ = ka + kQ(da/dQ) = ka + kQ(a/Q)$$

$$dK/dQ = 2ka \quad \text{(A7b)}$$

Thus, if reserves are fixed, the additional output is being transferred from future to present. But it takes twice as much investment to expand output by a given amount. Therefore, the net value of the additional barrel is less by the amount of development investment.

ACKNOWLEDGMENTS

The research for this paper has been supported by the National Science Foundation, grant No. SES–8412791, and by the Center for Energy Policy Research of the M.I.T. Energy Laboratory. I am obliged to Kevin Lam, Michael C. Lynch, Manoj Shahi, and Jeffrey Stewart for valuable assistance. Earlier versions were given at a seminar at Texas A&M University, and as an Olin Distinguished Lecturer at the Colorado School of Mines. I am indebted to the comments of James M. Griffin, John R. Moroney, and G. Campbell Watkins. But any opinion, findings, conclusions or recommendations expressed herein are those of the author, and do not necessarily reflect the views of the NSF or any other person or group.

This paper is a revision of 86-008WP.

NOTES

1. That is, ($70,000/365) * (.12 + .10)/.65 = $64.91.
2. During 1963–1985, this is the drilling cost index of the Independent Petroleum Association of America (IPAA 1963–1985). During the years of overlap 1963–1973, changes in the IPAA index were very close to those in the GNP:IPD price index of "non-residential gross domestic business investment." The two diverged sharply after 1973, an important fact to be discussed below. But it enables the use of the "business investment" index a proxy for the IPAA index for years before 1963.
3. In March 1986, the API resumed publication of a comparable series, starting with the year 1983.
4. During 1955–1982; designating oil development wells as OWLS, the cost index as IP, reported oil development expenditures could be fairly well estimated by:

$$\text{Expenditures (\$billions)} = \$7.6 * OWLS^{0.72} * IP^{1.03}$$

The standard error of estimate is 8.6 percent.

Incidentally, the number of wells, and the IPAA drilling equipping cost index, are reported promptly.

Using them to estimate expenditures normally saves one to two years.

5. Suppose all pools to be at the same depth, circular, and homogeneous. Then the total number of wells drilled will be proportional to area. The number of dry holes will be proportional to the circumference which they outline. The ratio of circumference to area is $2/r$, where r = radius, so the smaller the area the higher the dry-hole ratio. Increasingly heterogeneous reservoirs are in effect smaller circles, with higher dry-hole ratios. Noncircular areas have a higher ratio of circumference to area.
6. This paper received the accolade of its author being denounced by name by the then Governor of Texas, Mr. John B. Connally.
7. Reserves-added in 1986 were less than half of the 1983–1985 average, but more than the lows of 1977, 1979, or 1982, when real (inflation adjusted) oil prices were higher.

REFERENCES

Adelman, M.A. 1962. *The Supply and Price of Natural Gas.* Oxford, England: Basil Blackwell.

Adelman, M.A. 1967. "Trends in the Cost of Finding and Developing Crude Oil and Gas in the United States." In *Essays in Petroleum Economics*, edited by Gardner and Hanke. Boulder: Colorado School of Mines Press.

Adelman, M.A. 1990. "Mineral Depletion, with Special Reference to Petroleum." *Review of Economics and Statistics*, 72, 1.

Adelman, M.A., J.C. Houghton, G.M. Kaufman, and M.B. Zimmerman. 1983. *Resources in an Uncertain Future*. Cambridge, MA: Ballinger.

Adelman, M.A., H. DeSilva, and M.F. Koehn. 1989. "The Valuation of Oil Reserves 1946–86." MIT Energy Lab Working Paper No. MIT-EL 87–014WP (December 1987), Society of Petroleum Engineers SPE 18906 (March 1989).

Alberta. Energy Resources Conservation Board. 1984. *Reserve Report Series ERCB–18* (December 31).

American Association of Petroleum Geologists Bulletin. "World Energy Developments" (Annual publication).

American Petroleum Institute. 1959. *Petroleum Facts & Figures: Centennial Edition*.

American Petroleum Institute-American Gas Association. 1946–1979. *Reserves of Crude Oil, Natural Gas, and Natural Gas Liquids in the United States* (Annual publication).

American Petroleum Institute-Independent Petroleum Association of America-Mid-Continent Oil & Gas Association. 1955–date. *Joint Association Survey on Drilling Costs*.

Andersen, A., & Co. 1985. *Oil & Gas Reserve Disclosures: Survey of 375 Public Companies, 1980–1983*, Summary Edition.

Banks, F.E. 1985. "The Division of the Oil Market between OPEC and non-OPEC Countries." Paper presented at the *Seventh International Symposium on Petroleum Economics*, Laval University, Quebec, November 6.

Beninger, W.A., and D.C. Arndt. 1987. "Guidelines Can Improve Property-Acquisition Results." *Oil & Gas Journal*, 85, 41 (October 12): 39–45.

Bradley, H.B., ed. 1987. *Production Engineering Handbook*. Richardson, TX: Society of Petroleum Engineers.

Burns, A.F. 1934. *Production Trends in the United States since 1870*. New York: National Bureau of Economic Research.

Desprairies, P.C., X. Boy de la Tour, J.J. Lacour. 1985. "Progressive Mobilization of Oil Resources." *Energy Policy*, 13(December): 511–523.

Ebinger, C. 1985. "Market Stability: Worth Paying the Price." *OPEC Bulletin* (September 9–14), p. 19.

Franssen, H. 1985. "The World Needs Every Drop." *OECD Observer*, 135(July): 29–30.

Gruy, J.J., F.A. Garb, and J.W. Wood. 1982. "Determining the Value of Oil and Gas in the Ground." *World Oil*, 194, 4:(March): 105–108.

Hotelling, H. 1931. "The Economics of Exhaustible Resources." *Journal of Political Economy*, 39: 137–175.

Hubbert, M.K. 1962. *Report to the Committee on Natural Resources*. National Academy of Sciences Publication 1000–D. Washington, DC: U.S. Government Printing Office.

Independent Petroleum Association of America. 1963–1985. *Report of the Cost Study Committee*.

McCray, A.W. 1975 *Petroleum Evaluations and Economic Decisions*. Englewood Cliffs, NJ: Prentice-Hall.

Meyer, R.F., and M.L. Fleming. 1985. "Role of Small Oil and Gas Fields in the United

States." *The American Association of Petroleum Geologists Bulletin*, 69(November): 1950–1962.

Miller, M.H., and C.W. Upton. 1985. "A Test of the Hotelling Valuation Principle." *Journal of Political Economy*, 93(February): 1–25.

National Academy of Sciences, Panel on Statistics of Natural Gas. 1985. *Natural Gas Needs in a Changing Regulatory Environment*. Washington, DC: National Academy Press.

Nehring, R. 1978. *Giant Oil Fields and World Oil Resources*. Santa Monica, CA: The Rand Corporation.

Nehring, R. 1981. *The Discovery of Significant Oil and Gas Fields in the United States*. Santa Monica, CA: The Rand Corporation.

Oil & Gas Journal. 1959. (September 14), p. 67.

Oil & Gas Journal. 1965. Annual Review & Forecast Issue.

Oil & Gas Journal. 1980. (September 15), p. 57.

Oil & Gas Journal. 1985a. Annual Review & Forecast Issue.

Oil & Gas Journal. 1985b. (November 18), pp. 118, 120.

Oil & Gas Journal. 1985c. (November 25), p. 48.

Petrie, T.A., and S.W. Wright. 1985. *Frontier Petroleum Development Economics*. New York: First Boston Research.

Petzinger, T., Jr. 1987. "Texaco v. Pennzoil, Anatomy of a Jury's Deliberations." *Wall Street Journal* (May 8), p. 7.

Petroleum Intelligence Weekly. 1985. (November 18), p. 8.

Petroleum Intelligence Weekly. 1986. (November 6), p. 3.

Picchi, B. 1985. *The Valuation of Reserves*. New York: Salomon Brothers.

Ryan, J.M. 1965. "National Academy of Sciences Report on Energy Resources," *AAPG Bulletin*, p. 49.

Salomon Brothers. 1987. *Domestic Oils* (April 28), p. 8.

Salomon Brothers. 1988. *Petroleum Exploration and Production* (February 2).

Smith, J.L., and J.L. Paddock. 1984. "Regional Modelling of Oil Discovery and Production." *Energy Economics*, 6, 1: 5–13.

Uhler, R.S., with the collaboration of P.C. Eglington. 1986. *The Potential Supply of Crude Oil and Natural Gas Reserves in the Alberta Basin*. Ottawa: Economic Council of Canada.

U.S. Bureau of the Census. 1974–1982. *Annual Survey of Oil & Gas*. Washington DC: U.S. Government Printing Office.

U.S. Cabinet Task Force on Oil Import Control. 1970. *The Oil Import Question*, edited by G.P. Shultz. Washington, DC: U.S. Government Printing Office.

U.S. Energy Information Administration. 1977–date. *U.S. Crude Oil, Natural Gas, and Natural Gas Liquids Reserves*. Washington, DC: U.S. Government Printing Office.

Wall Street Journal. 1985. (January 1), p. 1.

Woods, T.A. 1985. "Resource Depletion and Lower 48 Oil and Gas Discovery Rates." *Oil & Gas Journal* (October 28).

THE PETROLEUM PROBLEM:
THE INCREASING NEED TO DEVELOP ALTERNATIVE TRANSPORTATION FUELS

D. J. Santini

ABSTRACT

An examination of the interactions of petroleum supply and demand patterns suggests that the depletion of U.S. oil reserves and the increasing importance of oil consumption in U.S. transportation are causes for greater attention to the development of alternative (nonpetroleum based) transportation fuels. Absent such development, the historical lessons in this paper suggest that another period of rising oil prices, erratic oil market behavior, and subsequent economic difficulty is probable within the next two decades. Methanol is argued to be the most likely and most desirable substitute transportation fuel, because it can be produced more economically and used more efficiently than gasoline when both are derived from the rapidly expanding worldwide supplies of natural gas.

I. INTRODUCTION

The basic argument of this paper is that it is imperative that the United States now devote increasing attention to developing substitutes for its petroleum-based transportation fuels. As Figure 1 shows, U.S. transportation now consumes more oil than is produced in the entire country. At the current level of oil prices, this situation is being exacerbated by a cutback in U.S. oil production and an increase in oil use in transportation. Contrary to popular opinion, the main cause of the post–1981 decline in oil prices and 1985–1986 oil price collapse was not increased fuel efficiency of vehicles. Instead, the primary cause was the adoption—in less than a decade—of substitutes for oil by electric utilities, industry, other businesses, and households. This conclusion is supported by data from Japan and Europe, as well as for the United States. Because transportation is using an increasing share of oil in the United States and other industrialized countries, the ability to implement substitute transportation fuels will be far more important when the next oil price run-up begins, possibly as early as the 1990s. In the case of the 1973–1981 oil price run-up, prices were not driven down until transportation belatedly began to reduce oil consumption, reinforcing the consumption drop already underway in other sectors. History shows that substitution in transportation fuel in the United States has typically taken decades (Santini, 1989a). If real oil prices are to be effectively capped below 1981 levels, the United States cannot risk waiting until after the start of another price run-up to begin implementing petroleum substitutes in transportation.

This paper examines the twentieth-century relationship of petroleum market behavior to economic behavior in the United States and the world. The history of U.S. petroleum markets and their past interactions with patterns of U.S. economic growth are discussed, and the implications of this history for transportation fuel markets through the end of the century are examined. More recent international trends in petroleum consumption and production are examined in detail, with special attention given to those multiyear periods when trends (i.e., growth rates) in domestic U.S. consumption and production temporarily diverge. These periods are referred to as "gaps." Such gaps are shown to precede sharp oil price movements. The oil price movements themselves are associated with periods of relatively severe U.S. economic difficulties. Sharp oil price increases—"supply shocks" in macroeconomic terms—theoretically (Dolan, 1986; McConnell, 1987; Kahn, 1987) and empirically (Santini,

1987a) are causes of recessions. Contrary to theory, twentieth-century statistical evidence and simple intuition, nineteenth- and twentieth-century U.S. history does not indicate that energy price collapses cause economic booms (Santini, 1987b). Thus, both recent experience and longer looks at historical evidence indicate that sharp energy price movements are not desirable.

This paper, and its supporting report (Santini, 1986), show that severe movements in petroleum price—both upward and downward—can be anticipated because they are preceded by a multiyear pronounced widening of the gaps just described. To reduce the severity of petroleum price swings, the gaps themselves logically must be reduced in magnitude. This paper argues that the petroleum market exhibits erratic price behavior because of difficulties in developing substitutes for oil. It is suggested that preparation of such substitutes must begin before price signals indicate the desirability of such preparation. If gaps can be "managed"—kept to a narrow range or substantially shortened in duration—the economy-wide consequences of subsequent wide swings in petroleum price should be avoidable. This, of course, is the same type of argument used by those who would have government intervene to reduce the inefficiencies of the business cycle (Dolan, 1986). However, the argument is limited in the sense that it applies only to one sector of the economy and unique in the sense that wide gaps are very early indicators for policy action, thus allowing reasoned, careful development of policies designed to narrow the gaps.

II. THE "PETROLEUM PROBLEM"

The "petroleum problem" faced by the United States and its allies has two facets: (a) the economic exhaustion of petroleum and (b) the increasing dominance of petroleum consumption by the fuel-inflexible transportation sector.

A. Economic Exhaustion of Petroleum

Undeniably, the United States is running out of economically recoverable petroleum. Further, it has been a stated goal of OPEC to keep world oil prices low enough to discourage further development of Alaskan and North Sea reserves (*Investor's Daily*, 1987). In and of itself, this is not a theoretical problem. Many nations have maintained robust economic growth despite a lack of indigenous petroleum. Exhaustion of domestic U.S. petroleum does, however, mean that the United States will no longer be able to close consumption-excess growth gaps by increasing domestic petroleum production. Therefore, the nation must rely increasingly on measures that:

- Increase exports to earn the necessary foreign exchange to purchase imported oil

and/or

- Improve the thermodynamic efficiency of equipment that uses petroleum products

and/or

- Develop domestically produced substitutes for petroleum products

and/or

- Import substitute fuels that are less costly.

B. Increasing Dominance of Petroleum Consumption by the Fuel-Inflexible Transportation Sector

The ability of the United States and its allies to rapidly reduce consumption of petroleum products is diminishing because the transportation sector has increased its share of the industrialized nations' petroleum products consumption (Figures 1, 2) (American Petroleum Institute, various issues; U.S. Department of Energy, various issues; Dunstan, et al., 1938; Organization for Economic Co-operation and Development, 1984; U.S. Department of Commerce, 1975; *Energy Statistics Sourcebook*, 1987). Not only does transportation account for a historically high share of U.S. oil consumption, it also accounts for a far greater share of petroleum consumption than that in competing industrialized nations (Figure 2). Further, among member nations of the Organization for Economic Cooperation and Development (OECD), the United States is far and away the most dominant user of oil in transportation (Figure 3). This is a problem because each of the other petroleum-using sectors of the U.S. and other OECD economies managed to reduce petroleum consumption earlier and more rapidly than did transportation after the two 1970s oil price shocks (Table 1) (Organization for

The Petroleum Problem 63

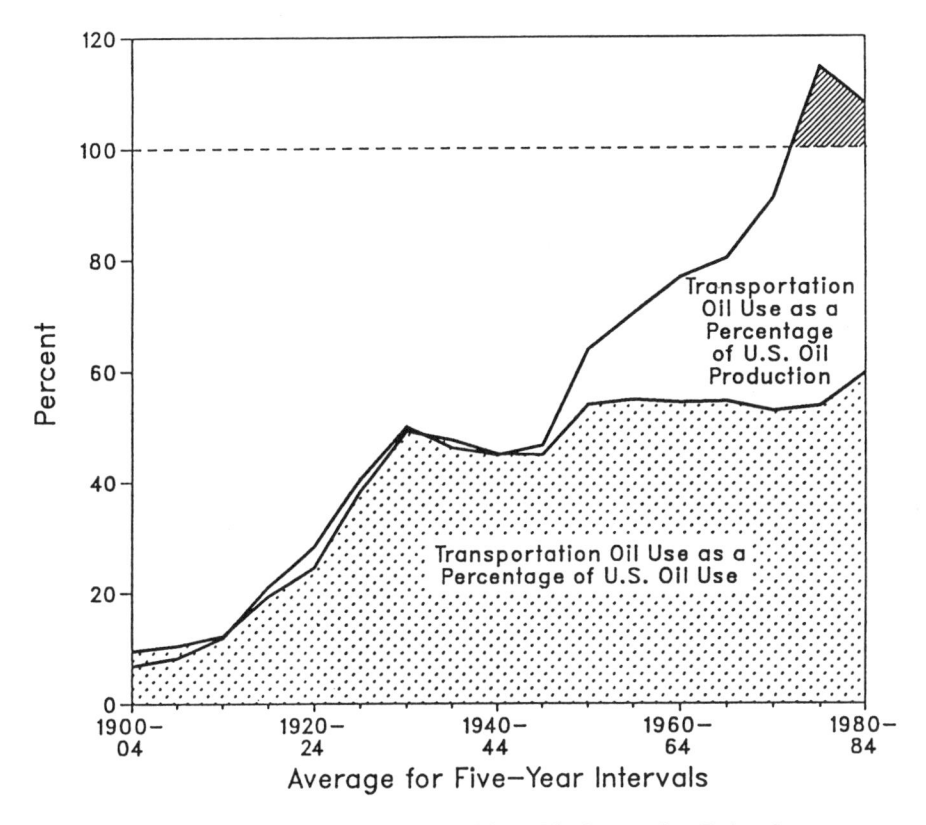

Figure 1. The Petroleum Problem: The Increasing Role of Transportation, 1900–1984

Economic Cooperation and Development, 1984). In spite of legislation that mandated rapid gains in automotive fuel efficiency in the United States, transportation in the United States had the least success in reducing petroleum consumption through 1985 (United States Department of Energy various issues). This pattern also held in Japan and Europe (Organization for Economic Cooperation and Development, 1984; *Energy Statistics Sourcebook*, 1987).

C. Oil Prices, Oil Imports, and Macroeconomic Activity

On another occasion in the United States (1915–1920), there was a doubling in real crude oil price during a period accompanied by increasing oil imports and immediately followed by a severe recession (Santini,

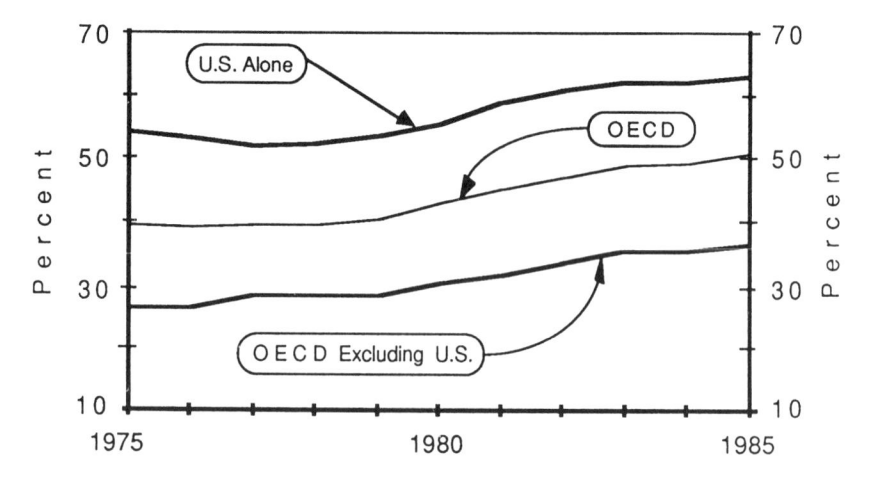

Figure 2. Shares of Oil Consumed by Transportation in the United States and Other OECD Member Nations, 1975–1985

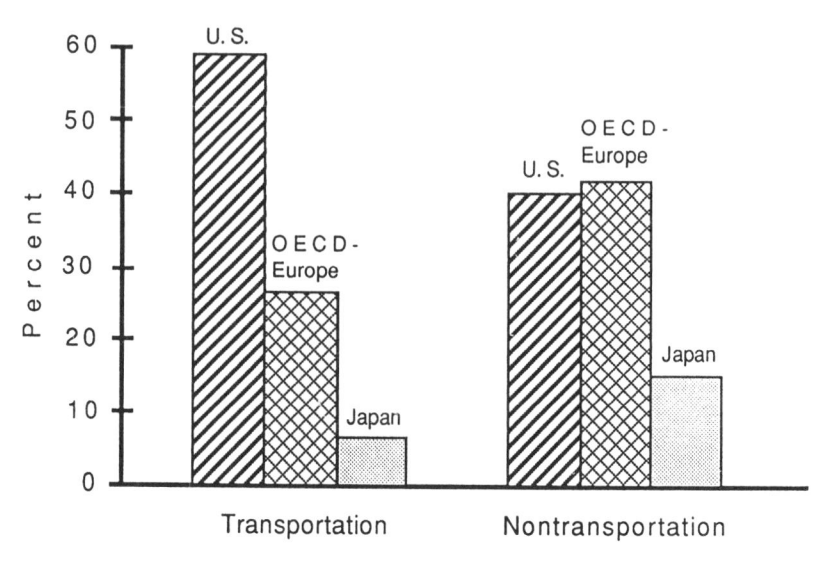

Figure 3. Shares of Total OECD Petroleum Product Consumption Accounted for in 1981, by Sector and Location

Table 1. Percentage Decline in Oil Consumption from Peak Year through 1981 for Major Oil-Consuming Locations, by Sector, Post-1978–1979 Oil Price Shock[a]

Location	Industry Decline	Peak Year	Residential, Commercial, and Other Oil Consumers Decline	Peak Year	Electric Utilities Decline	Peak Year	Transportation Decline	Peak Year	Overall Decline	Peak Year
United States	23.1	1979	20.4	1977	43.4	1978	8.4	1978	13.3	1978
United Kingdom	28.5	1979	27.3	1979	50.8	1978	3.5	1980	16.6	1979
West Germany	21.5	1979	24.3	1979	35.6	1978	3.1	1980	18.2	1979
OECD-Europe	25.5	1979	16.3	1978	16.7	1978	1.6	1980	14.8	1978
Japan	28.4	1979	0.6	1980	25.8	1977	13.2	1979	22.2	1979

Note: [a]For year in which oil consumption peaked through 1981.

1986; U.S. Department of Commerce, 1975). Although macroeconomic theory now recognizes the oil "supply shocks" as causes of recent recessions (Dolan, 1986; McConnell, 1987; Kahn, 1987), very little reexamination of the role of such supply shocks as causes of pre-World War II recessions has taken place. In this paper we make the assumption that if oil price shocks can be cited as causes of recent recessions in macroeconomics texts, other similar oil price shocks may also be assumed to have causally contributed to earlier U.S. recessions if statistical analysis supports such an assumption. Separate statistical analyses of the pre- and post-World War II periods do support this position (Santini, 1987a).

Thus, if a recession-inducing crude oil price increase (Santini, 1987a) comparable to that of 1915–1920 or 1978–1981 (real price increases more than doubled in both cases; (see Santini, 1986; American Petroleum Institute, various issues; U.S. Department of Commerce, 1975) occurs in the future, the United States—all other things being equal—would be less able to reduce oil import purchases and costs than in the past. The nation would also suffer more than its industrial competitors, who should have less difficulty reducing their oil consumption because the inflexibility of the transportation sector is relatively less of a problem for them (Figure 3). This appears to have been the case in the years immediately after the large 1978–1979 oil price jump, because Japan and Germany achieved greater immediate success than the United States in reducing oil consumption (Table 1). Similarly, since the first oil price shock in 1973, total oil consumption as of 1986 had dropped by 19.1

percent in Japan, 17.7 percent in West Germany, only 7.8 percent in the United States, and 16.5 percent elsewhere in the OECD (*Energy Statistics Sourcebook*, 1987). These declines in total oil consumption occurred in spite of an 11 percent increase in U.S. oil consumption in transportation from 1975–1985 and an even greater 23 percent increase over the same period in non-U.S. OECD countries! (U.S. Department of Energy, various issues; *Energy Statistics Sourcebook*, 1987)

The purpose of this paper is to note and emphasize the possible importance of these facts in order to support the argument for the increasing need for well-conceived transportation strategies that can quickly reduce petroleum consumption. The paper is not intended to prove that described energy–to–macroeconomy relationships are correct. It is intended to raise issues and provoke thinking about the importance of advanced planning for quick introduction of oil substitutes when oil prices rise.

III. MANAGING THE NEXT GAP

A. The Need in Transportation for a Petroleum Substitute

Implementation of substitutes in transportation will prove to be one of the most difficult transitions ever to occur within the U.S. economy. Because of their physical properties, petroleum-based fuels make ideal transportation fuels. Potential substitutes are costly and often inferior in key respects. Nevertheless, significant use of substitutes will eventually be needed, given the inevitable depletion of economical oil resources. Because the United States pioneered in the extensive use of oil in transportation (Dunstan et al., 1938; Williamson and Daum, 1959), it should logically be one of the first, if not the first, to face the need for developing a widely used substitute.

B. Methanol: A Possible Transportation Fuel Based on Natural Gas

This paper briefly considers methanol, the fuel that many current analysts find to be the most likely widespread substitute for petroleum-based gasoline (Sobey, 1989; Gray and Alson, 1989; Gray and Alson, 1985; Swedish Motor Fuel Technology Co., 1986; Society of Automotive Engineers, Inc., 1990; Mossavar–Rahmani, 1986; McNutt and Ecklund, 1986). Methanol is a liquid fuel suitable for automobiles and can be made more cheaply from natural gas than can gasoline (Swedish Motor Fuel Technology Co., 1986). While worldwide discovery rates of crude oil through 1987 had been disappointing, natural gas added relatively rapidly to world reserves (American Petroleum Institute; *Energy Statistics Sourcebook*, 1987). Drilling for oil has taken oil companies to increasingly remote locations and greater depths. However, at greater depths the probability is higher that gas rather than oil will be found (Mossavar-Rahmani, 1986). Further, gas price increases of the 1970s caused increases in the search for natural gas for its own sake. Previously, most natural gas was an unwanted product associated with oil recovery. Recent changes in drilling patterns have resulted in far greater success in finding natural gas than oil. From the oil price shock of 1973 through 1986, world crude oil reserves increased by 5 percent, while gas reserves increased by 93 percent (*Energy Statistics Sourcebook*, 1987). Thus, as oil becomes increasingly more difficult to find, the issue will be the kind of transportation fuel to make from the more readily available supplies of natural gas.

Automakers, the U.S. Environmental Protection Agency (EPA), the State of California, and numerous other entities are spending significant amounts of money on experiments to assess and improve the ability of U.S. automobiles to burn methanol cleanly and efficiently. The degree of success of these and other experiments and the timing of future market development efforts will have a major effect on when or whether the world and the United States again experience another multi-year crude oil price run-up as high as those of 1915–1920 or 1978–1981.

IV. THE SLOW PROCESS OF REVERSAL OF OIL PRICE INCREASES

A. A Comparison of the 1915–1931 and 1968–1986 Oil Price Paths

The two worst twentieth-century run-ups of crude oil price in the United States occurred in 1915–1920 and 1978–1981, each after one of the two most prolonged and pronounced gaps in which growth rates of oil consumption exceeded those of domestic production. When these gaps were reversed and replaced by the twentieth century's most prolonged and pronounced gaps of the opposite type — in which rates of production growth exceeded rates of consumption growth — oil prices were driven back down (Figure 4). This strong reversal of the consumption-vs.-production-growth gaps was due in part to unusually severe declines in real national income, although energy conservation and fuel substitution also played an important role. Fifteen years after the steady real-price rise that began in 1915, oil prices plummeted to their 1915 level. In 1986, 13 years after a price rise that began in 1973, oil prices once more plummeted toward their preshock level (Figure 5). Dunkerley and Hoch (1986) used international data to estimate that the price elasticity of "road transport" oil consumption is very low (−0.2), but that income elasticity is high (+1.3); thus, I infer that the income losses in recessions and larger price responses in other sectors do more to drive oil prices back down than do price responses by the transportation sector. One major purpose of this paper is to decry this unfortunate inference and to argue, like Dunkerley and Hoch, that actions to improve the price response are desirable if governments are to avoid using income losses to push down oil prices after price shocks (Dunkerley and Hoch, 1986).

That unusually severe U.S. recessions followed both of the worst U.S. real oil-price run-ups during both of these periods implies that balance in the energy sector must be maintained if economic stability is to be achieved. If a prolonged and broad consumption-excess growth gap emerges, history implies that such a gap must be balanced by another prolonged and broad gap in which growth rates of production exceed those of consumption. During the twentieth century under the U.S. economic system, this has twice been accomplished by an initially sharp, and then prolonged, increase in real crude oil price that encouraged production, discouraged consumption, and — consistent with the "supply shock" theory (Dolan, 1986; McConnell, 1987) — caused a reduction of national output far below the "natural" level. Consistent with the theoretical recovery process from a supply shock, these combined effects reversed the gap and depressed the crude oil price. On the first occasion (1915–1923), this process was followed by a period when consumption and production were more or less in balance, and, aside from one mild recession, steady, robust economic growth (the "Roaring Twenties") was experienced from the recession trough in 1921 to the end of the decade. In that case, we had the benefit of domestic oil discoveries but later paid the costs of writing down the value of those oil assets in 1931, during the 1929–1933 collapse into the Great Depression.

On the first occasion, an initial sharp oil price run-up started in 1915

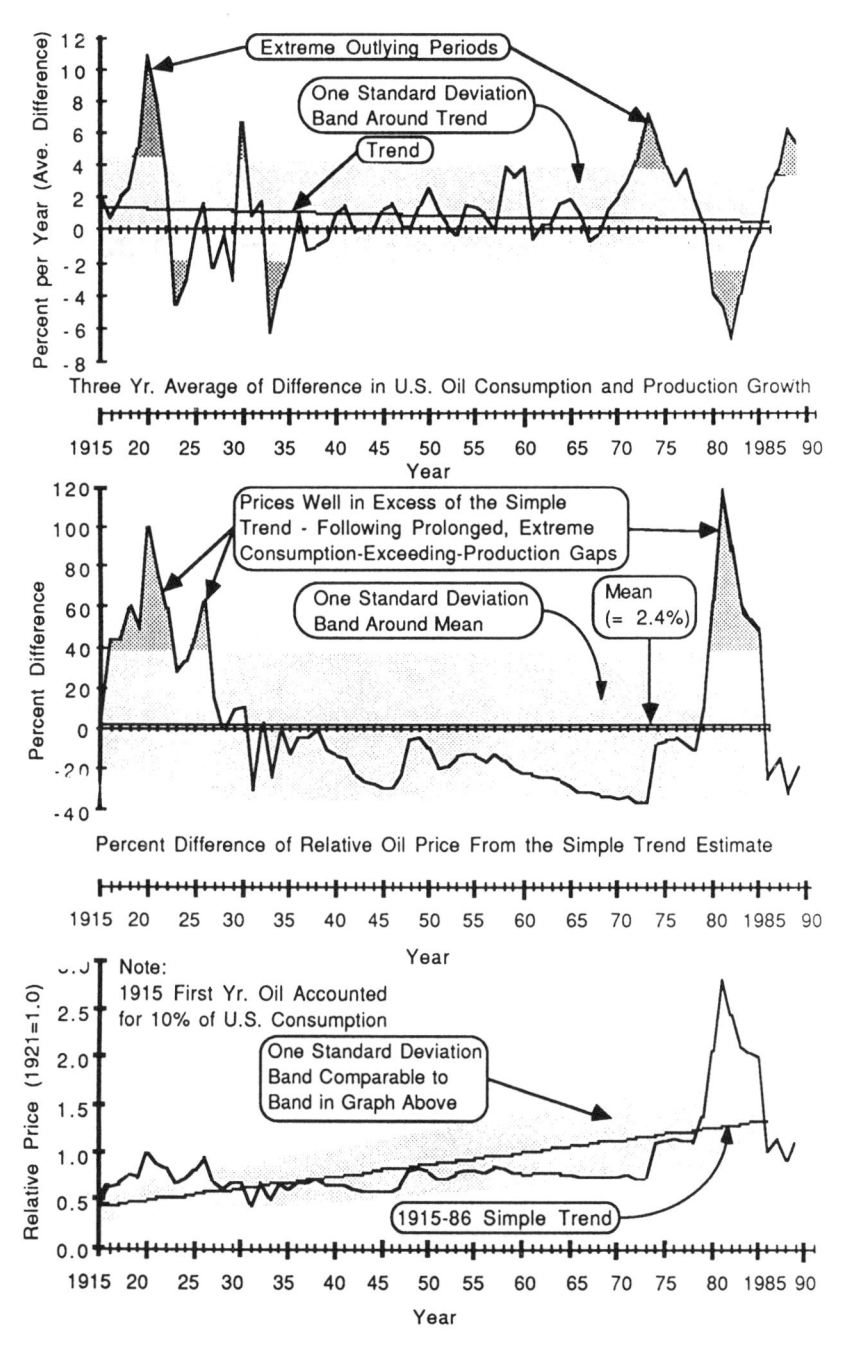

Figure 4. Movements of U.S. Oil Consumption-vs.-Production-Growth Gaps against Real Oil Prices, 1915–1986

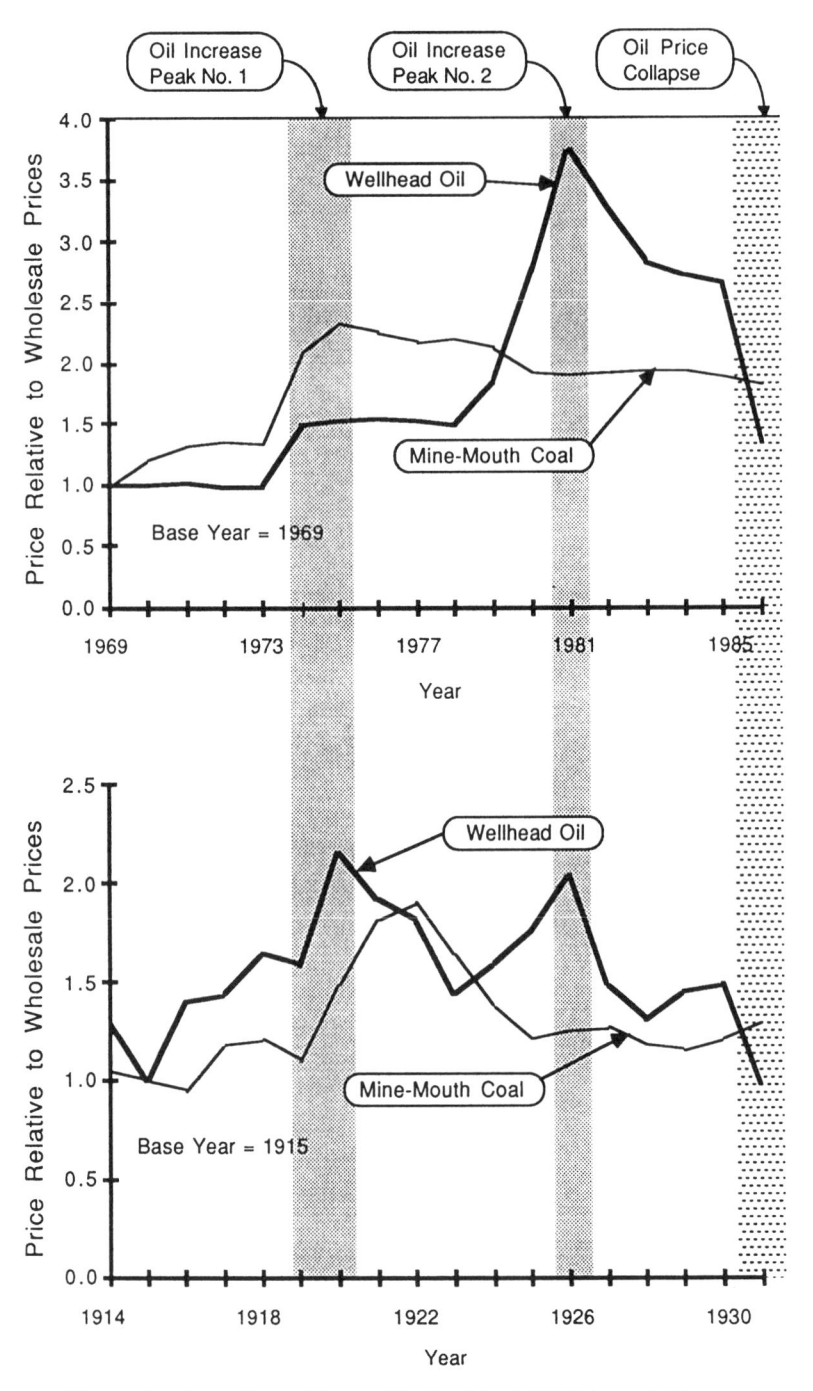

Figure 5. Long-Term Rise and Fall of Real Oil Prices: A Comparison of the 1915–1931 and 1973–1986 Periods

The Petroleum Problem

and was followed by a second oil price shock and subsequent collapse 15 years later. On the second occasion (1968–1986), we did not find much new domestic oil after the first price shock, and a second shock — more severe than that of the mid-1920s — ensued. On this occasion, oil and transportation accounted for more of our total energy consumption than in the 1915–1931 case, and the start-to-peak real oil price increase was greater (Figure 5). When oil prices collapsed in 1986, the absence of domestic discoveries was an advantage in that other nations carried a greater part of the burden for writing off oil assets. Nevertheless, recent events in the financial markets raise the possibility that the process of writing off domestic oil assets may once again have contributed to financial difficulties in the banking system. Low oil prices were, in 1988 reflected in low stock values for energy-center banks in particular, and the banking system in general, as well as record post-World War II bank failures. The oil market's behavior in 1916–1931 and 1968–1986 is similar in several respects. Both saw dramatic real oil price run-ups, with two distinctly different price shocks peaking six years apart. After the two price shocks, both periods witnessed prolonged and severe erosion of crude oil prices. In both cases, a collapse in crude oil prices occurred about five to six years after the second oil price shock (see Figure 5). Subsequent events, as of 1990 also implied that a rebound from the collapse and return to pre-collapse real oil price levels could also be a similarity of the two periods (see Figure 4).

A key finding of this analysis is that price movements within the petroleum market balance out (come to equilibrium), as basic economic theory would lead one to expect (Dolan, 1986; McConnell, 1987). However, the process by which balance is reestablished (after a consumption-excess growth boom) has historically been very long in duration and has involved recessions after the upward price movements. The analysis thus implies that the initial gaps themselves precede price shocks that cause subsequent declines in economic activity. When the gaps are created through declines in crude oil production, this sequence of events is consistent with "supply shock" textbook theory (Dolan, 1986; McConnell, 1987). The widening of the gaps that precedes the price shocks seem to be identifiable well in advance of the shock (including the 1988–1990 oil price run-up) and national income declines that follow (Figure 4). Given these characteristics, it seems both possible and desirable to take steps to reduce the size of gaps before they contribute to the consequences identified here. Thus, it is argued that the causes of gaps should be considered in advance so that the size of gaps can be limited — that is, gaps should be "managed."

B. Statistical Evidence on Sectoral Responses to Oil Prices from 1973 to 1986

The descriptive examination of the 1915–1931 and 1968–1986 periods showed that a production–in–excess–of–consumption growth gap eventually drove real oil prices down in response to two prior large real oil price increases that were separated by about six years. Prices were not pushed back to the level existing at the beginning of each period until this production–in–excess–of–consumption growth gap emerged. One question that arises from observation of this pattern of market behavior is "why was the initial price rise insufficient to cause enough substitution to drive the relative price back to its original level?" Although this paper emphasizes the relative difficulty of introducing substitutes for petroleum in transportation, it is also true that it is difficult, in an absolute sense, to introduce substitutes for petroleum in the short run in any major petroleum-using sector of the economy.

The process of substitution can be described as occurring in two phases logically separated by a few years. First, immediately after a price shock, consumers and investors substitute among existing technologies to reduce oil consumption as much as possible. Some industrial and business consumers take advantage of a built-in ability to immediately switch fuels. In the second part of the substitution process, producers of energy-intensive products begin programs of research, development, and product introduction to implement new technologies that are more thermodynamically efficient and/or use more abundant, less expensive fuels. The lag between the start of such programs and the first commercial sale of new products is typically several years. For example, flow charts (Amdall, 1987; Sekar and Tozzi, 1983; Rossi, 1986) and public statements by manufacturers indicate that five years are needed to complete engine development programs. If such lags also exist for furnaces, air conditioners, boilers, electric generators, and process equipment, it might be reasonable to expect a lagged reaction to sharp oil price increases after about half a decade. If this description of the substitution process is correct, then on those occasions when the first round of substitution — using existing products — is sufficient to push the price down to its original level, product development programs might well be shelved in

The Petroleum Problem

favor of continuing production of existing technologies. However, if the initial substitution among commercially available technologies does not push prices back down, a second price shock could be expected to cause commercial introduction of the new technologies. If this is an accurate description of the process, the combined short- and long-term substitution responses to the pair of price shocks should be sufficient to push oil consumption below production and induce the needed price decline.

The problem with this sequence is that it implies that several years of sustained high prices must occur to cause the widespread adoption of new oil-substituting and oil-conserving technology. A.J. Sobey of General Motors has argued that it would take several years of gasoline prices well above recent levels to induce widespread substitution of methanol for gasoline (Sobey, 1989). If the process just described is accurately depicted, it is legitimate to expect the temporary but severe economic difficulties associated with petroleum price shocks to once more occur — if significant development and implementation of oil-substituting technology in the transportation sector is not begun until after another substantial rise in oil prices.

To test for the existence of these effects and to statistically describe the nature of this process, the sector-specific short- and long-run price elasticities of U.S. oil (and energy) consumption rates per unit of output for the 1973–1986 period were estimated using an individual, 13-observation ordinary least squares regression model for each of the four major petroleum-consuming sectors, as categorized by the Energy Information Administration. The oil consumption data series that was used begins with 1973 (U.S. Department of Energy, various issues). Annual oil consumption per unit of output was estimated by dividing sector-specific oil consumption by a measure of output for that sector. The output index for electric utilities was net generation (U.S. Department of Energy, various issues), for industry it was the Federal Reserve Board industrial output index (U.S. Bureau of the Census, various issues), for residential and commercial oil consumption it was constant-dollar personal consumption expenditures (U.S. Department of Commerce, 1987), and for transportation it was vehicle-miles of travel, or VMT (U.S. Department of Transportation, 1986). The advantage of measuring changes in energy use per unit of output is that the short-run elasticity estimates do not suffer from the upward bias (of the absolute magnitude of the negative elasticity estimate) resulting from the output losses that immediately follow an energy price shock. This short-run bias showed up in unreported regression models in which I tested for its existence.

In the case of the residential/commercial sector, it was reasoned that the best measure of "output" would be a proxy for consumer utility, and real personal consumption expenditures were selected as the best proxy measure. In the case of transportation, the VMT series ended in 1985, so transportation estimates do not include the immediate response to the 1986 oil price collapse. A check of the effect of deleting 1986 in the other models of oil consumption showed no sign changes in otherwise significant variables, as well as consistently larger changes of elasticity estimates for short-run effects than for long-run effects. Changes in short-run elasticity estimates were −26 to +80 percent, while changes in long-run estimates were −11 to +18 percent. The largest coefficient shifts occurred in the residential/commercial equation, where the output index is most questionable in any case. On the whole, given the rather large differences in elasticity estimates across sectors, these experiments suggest that the transportation estimates can be reliably compared to those for the other sectors in spite of the exclusion of the "outlying" 1986 observation in the transportation model.

The rate of change of oil consumption per unit output (O_0), rate of change of energy consumption per unit (E_0), and of real crude oil price (P_i—in 1967 dollars) (American Petroleum Institute, various issues) were computed by logging (using the base 10) the ratio of this year's value to last year's value. Short-run elasticities were initially estimated by regressing this year's O and E values against present (P_0), and past oil price changes ($P_1, P_2, \ldots P_i$). Statistically significant price-induced reductions in O occurred immediately in the electric utility, industrial, and residential/commercial sectors, but not until the year after the oil shock for transportation (see Table 2). After a lag of one year, statistically significant effects consistently disappeared in regression models with two or more P_i lags included. For comparability across all sectors, elasticities for contemporaneous reactions and for reactions after one year were estimated for every equation. For short-run elasticities, the remaining discussion refers to the sum of the P_0 and P_1 coefficients (elasticities) for each sector's equation. These sums are reported in Table 2, along with the separate P_i coefficients.

Long-run elasticities were estimated by an iterative procedure. The independent variable measuring long-run oil price pressure (L) used the log of the following quantity: oil price in a given year divided by oil price

Table 2. Sectoral Estimates of Short-Run and Long-Run Elasticity of Oil and Energy Consumption per Unit of Output with Respect to Oil Price[a]

	Short-run P_0	P_1	2-yr Sum $(P_0 + P_1)$	Long-run (L)	Adjusted R^2
Oil Sector					
Electricity	−0.424 (−3.71)[b]	−0.008 (−0.05)	−0.432	−0.462 (−3.74)[b]	0.665 (8.91)[b]
Industrial	−0.149 (−2.69)[c]	−0.050 (−0.66)	−0.199	−0.158 (−2.96)[b]	0.428 (3.97)c
Residential/ Commercial	−0.123 −1.85[c]	−0.126 (−1.36)	−0.249	−0.073 (−1.22)	0.242 (2.27)
Transportation	−0.010 (−0.57)	−0.042 (−2.58)[c]	−0.052	−0.045 (−3.21)[b]	0.677 (8.70)[b]
Energy Sector					
Electricity	0.002 (0.22)	0.001 (0.05)	0.003	−0.003 (−0.36)	−0.294[d] (0.09)
Industrial	−0.025 (−1.07)	−0.005 (−0.16)	−0.030	−0.072 (−3.14)[b]	0.379 (3.43)[c]
Residential/ Commercial	~0 NE[e]	~0 NE	~0 NE	−0.023 (−1.65)	−0.065[d] (2.73)
Transportation	0.004 (0.21)	−0.047 (−2.48)[c]	−0.043	−0.038 (−2.31)[c]	0.584 (6.15)[c]

Notes: [a] t-statistics for coefficients are shown below the coefficient in parentheses; F-statistics for model R^2 are below the R^2 value in parentheses.
[b] Statistically significant at the 1 percent level (two-tail).
[c] Statistically significant at the 10 percent level (two-tail).
[d] A property of the small sample adjustment of the R^2 statistic is that a negative value can result from the adjustment.
[e] NE = Not estimated in the "best" model presented in the table; approximately zero in other estimated equations.

in an earlier year. The interval between the initial year and earlier year was selected by examining R^2 improvements and coefficient t-values when the long-run variable was added to equations estimating short-run elasticities. Intervals as long as seven and as short as two years were tested. The best interval proved to be six years. This interval was used throughout in the reported regression results. In addition to this iterative "optimization" procedure, the long-run price variable itself was lagged from zero to three years ($L_0, L_1, \ldots L_3$) and a "best" lag was determined, also based on R^2 improvements and coefficient t-values. These lags were one year each for electricity and transportation, two years for industry, and three years for residential/commercial. The elasticity coefficient for

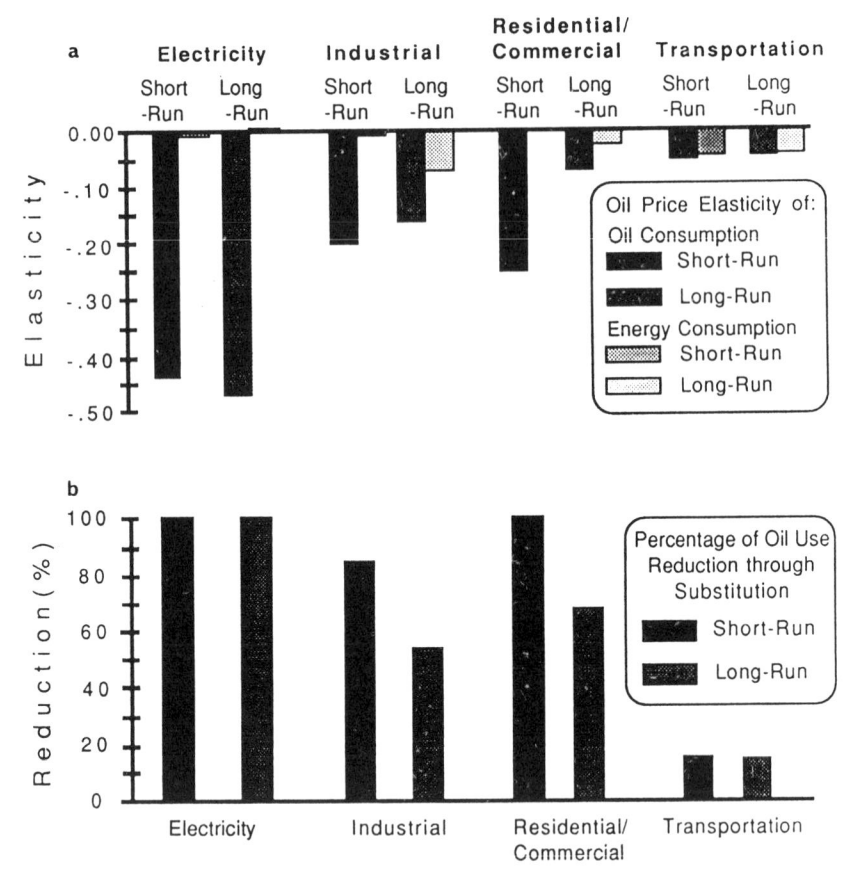

Figure 6. Sectoral Estimates of Short- and Long-Run Elasticity of Oil and Energy Consumption in Response to Real Oil Price: A Comparison to Estimates of Substitutability for Oil, 1973–1986

this long-run price variable indicates the longer-run effect of a quasipermanent price increase, rather than the short-run shock effect indicated by the *P* coefficients. Other long-term lag formulations, such as distributed lags, have not been tested at this time. Results of these experiments are reported in Table 2.

Each regression model included a constant term, which is not reported. Short- and long-run elasticity estimates by sector are illustrated in Figure 6. The degree to which substitution versus straight conservation is responsible for reduced oil consumption in a given sector can be approximated by comparing short- and long-run elasticities for total energy consumption to those for oil-derived energy consumption. The ratio of the oil price elasticity of total energy consumption to that for oil energy consumption (times 100) approximates the percentage of the oil–energy–per–unit–of–output (O) reduction achieved by conservation. Subtracting this quantity from 100 approximates the percentage achieved through substitution. The results of this latter substitution share approximation are shown at the bottom of Figure 6, directly under the elasticity values from which they were derived.

The number of observations in these regressions is relatively small — lower in fact than found in statistical tables for the Durbin–Watson d-statistic. Based on linear extrapolation of the d-values, application of the Durbin–Watson test (results not reported) did not allow conclusive determination of the presence or absence of autocorrelation. Because of the small number of observations, the R^2 value presented is the adjusted value (Merrill and Fox, 1970).

The statistical results of these experiments tend to confirm this paper's descriptive analysis of the relative difficulty of substituting for oil in transportation. The estimates for both short- and long-run elasticity for transportation are substantially smaller in absolute magnitude than for the other sectors (Table 2 and Figure 6). Further, supporting the earlier descriptive arguments, the percentage of reduction in oil use estimated to have occurred through fuel substitution is far lower for transportation than for any other sector (Figure 6). These results also support the argument that there is a very long lagged technological response to sustained energy price increases, which takes hold only after years of effort to research, develop, and introduce new and improved energy conversion technologies.

V. LONG-TERM LIMITATIONS OF U.S. OIL SUPPLIES

A fundamental problem faced by the United States is the declining domestic supply of economically recoverable oil (Figure 7). Since 1900, the growth rates of oil consumption and production have moved in a clear, although slightly erratic, downward path. Ominously, in the 1970s, domestic U.S. production "growth" rates turned negative. While consumption growth rates have not declined as rapidly as production growth rates, it is clear that consumption growth rates have steadily declined in concert with domestic production limitations. The unsteady reduction of

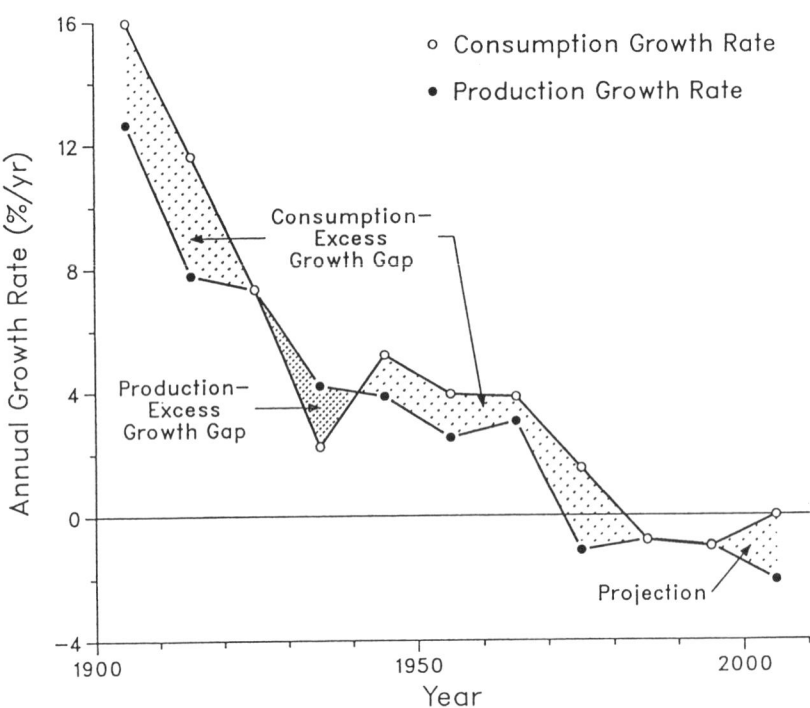

Figure 7. Decade Averages of Oil Consumption and Production Growth Rates (History to 1980, Projections to 2005)

consumption growth rates until they shifted into decline in the early 1980s was promoted by unsteady but clear long-term increases in oil price relative to prices of other goods (Figure 4). Domestic production of oil is clearly constrained by cost. The unprecedented price increases of crude oil of the 1970s only reduced the rate of decline of domestic production, rather than leading to an actual increase in domestic production. Rates of consumption growth, however, were turned into declines even greater than those in production. Thus, the production-excess growth gap of the early 1980s was far different than that of the early 1930s (Figures 4 and 7).

Because growth rates of oil consumption were consistently above those of production during 1945–1978 (Figure 4), imports were needed to make up the difference (Figure 8). On average, the difference between consumption growth and production growth was relatively steady and slight during 1950–1968 (Figures 4 and 9), leading to a slow increase in the gross percentage of imported oil from about 13 percent to 21 percent.

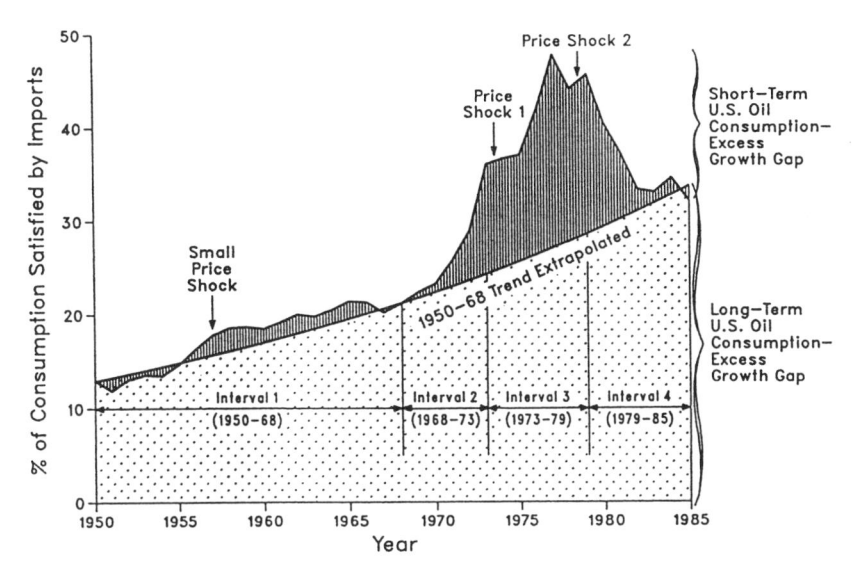

Figure 8. Short- and Long-Term Patterns of Gross Oil Import Shares in Domestic U.S. Oil Consumption, 1950–1984 (Source: Refs. 1,8)

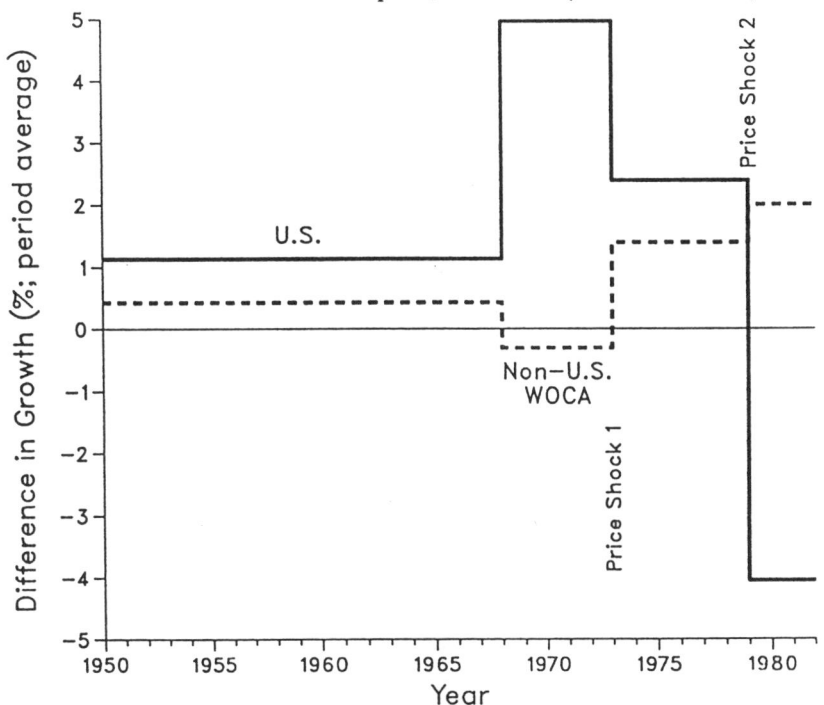

Figure 9. Difference between Growth Rates of Oil Consumption and Production for U.S. and World Outside Communist Areas (WOCA), 1950–1982 (period averages; source: Ref. 1)

The United States found that imported oil was cheap, and thus it maintained trade surpluses throughout the period; through 1968 this trend was readily manageable. Given the growth trends of oil price and domestic consumption and production for the period, this pattern probably could have continued for many years.

VI. CAUSES OF THE 1968–1977 GAP

A. Regulations

The steady trend of 1950–1968 did not continue. Legislation was passed to improve safety in underground coal mines, most of which were in high-sulfur eastern coal fields. At the same time, sulfur dioxide emissions standards for power plants and industry were legislated and promulgated. The result was a sharp increase in the cost of burning the dominant coal, that is, high-sulfur eastern bituminous. This was reflected in a coal price shock during 1968–1971. Consistent with supply shock

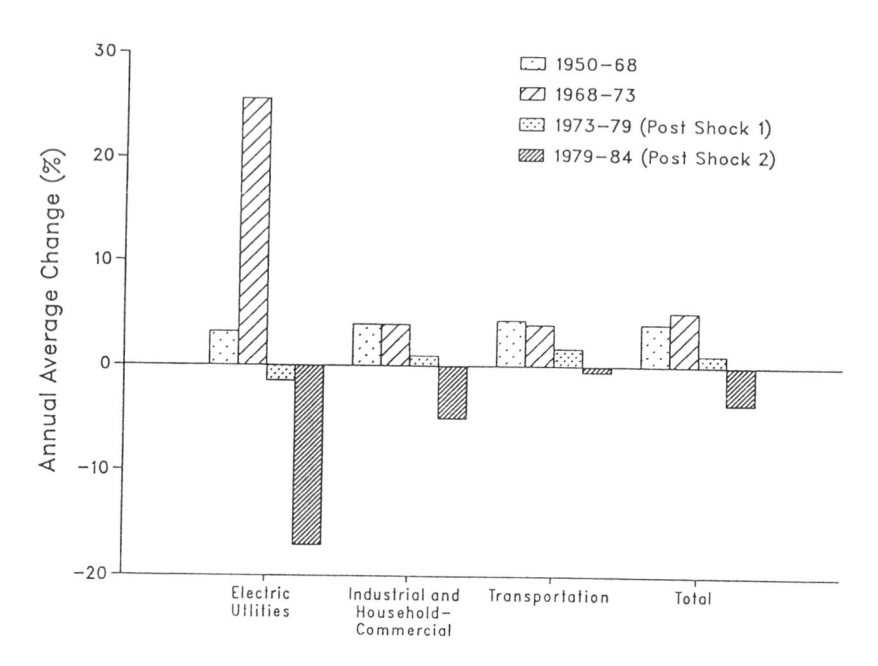

Figure 10. Changes in Sectoral Rates of Growth of Oil Consumption, 1950–1984 (period averages)

The Petroleum Problem

theory, a recession occurred in 1970. Unable to burn coal cleanly and economically under the new regulations, electric utilities and industry began a massive switch to oil during 1968–1972 (Figure 10). This led to an average increase of more than 1 percent/year in the growth rate of total oil consumption during 1968–1973.

B. Absence of Transportation Influences

The 1968–1977 gap was not, contrary to popular opinion, created by events in the transportation sector, where the growth rate of oil consumption actually dropped below the 1950–1968 average (Figure 10). This gap led to a sharp increase in the need for, and use of, imported oil (Figure 8) — especially from the Middle East — making the United States highly susceptible to the imposition of an oil embargo. The ensuing crude oil price rise (Figure 4) cannot, according to the logic in this report, be attributed to anything other than the U.S. gap, given the absence of such a gap worldwide (Figure 9).

C. Oil Production

Production of U.S. oil peaked during 1968–1973 just as consumption growth began its regulation-induced increase. The combination of increasing consumption growth and decreasing production growth created a severe consumption-excess growth gap in the United States (Figure 9) at the same time that free-world consumption growth rates were actually below production growth rates.

D. Oil Imports

When growth of oil consumption exceeds that of domestic oil production for a period of years, it is obvious that the extra oil must come from elsewhere. The importing of oil allows the gap to emerge, but also makes the U.S. oil supply and economy more susceptible to events beyond its borders. Each of the most severe U.S. oil price run-ups in the twentieth century occurred when oil import percentages were at all-time highs (Figure 11). It appears that in both cases the market decided that the increasing dependence on imports had to be reversed, because the ensuing oil price increases led to subsequent sharp drops in the percentage of imported oil.

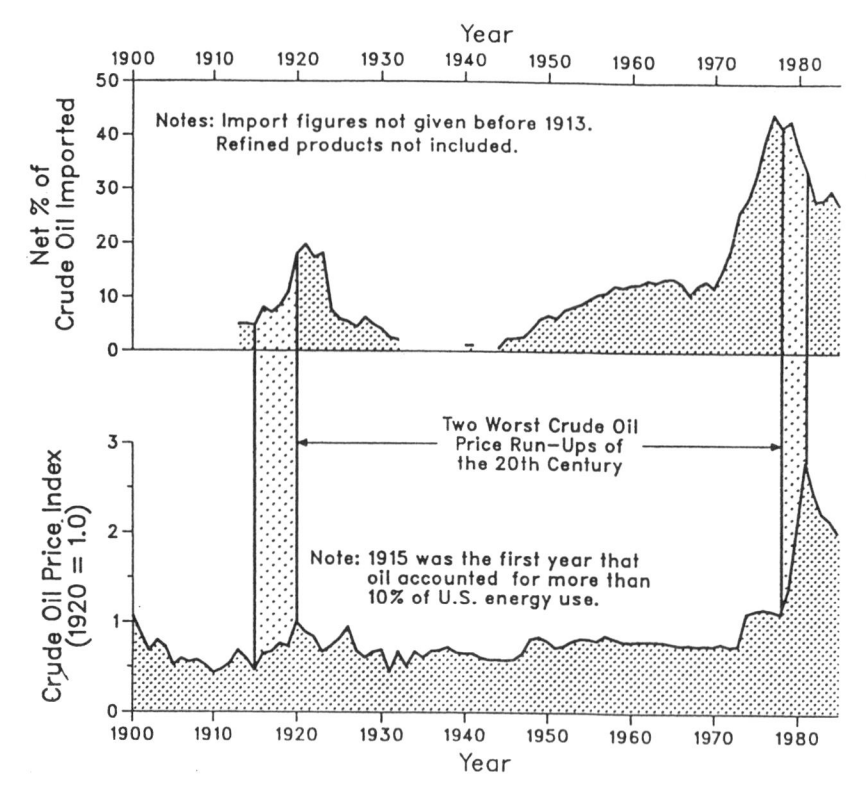

Figure 11. Comparison of Net Percentages of Imported Crude Oil to a Crude Oil Price Index, 1913–1982

VII. INCREASING NEED TO DEVELOP OIL SUBSTITUTES IN TRANSPORTATION

A. Role of Transportation in Overall Oil Consumption

As mentioned, transportation was less responsible for initiating the 1968–1979 gap than were other sectors (Figure 10). While the other sectors did increase their oil consumption during 1968–1973, they were able to rapidly reverse these increases so that they contributed far more to the closure and reversal of the gap than did transportation. This was possible mainly because it was far easier for these sectors to switch fuels than it was for transportation. In fact, transportation's growth momentum and its inflexibility in fuel substitution and conservation caused this

The Petroleum Problem 83

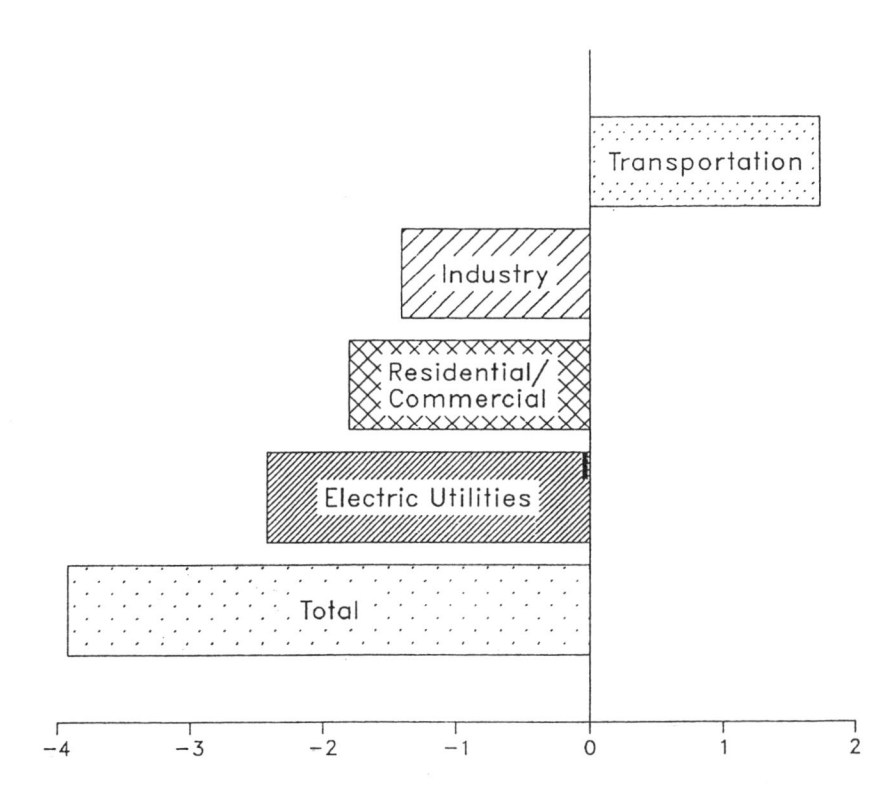

Figure 12. U.S. Sectoral Changes in Oil Consumed per Year, 1985 vs. 1973 (in quads; 1 quad = quadrillion, or 10^{15}, Btu)

sector to consume more oil in 1985 than in 1973, while each of the other sectors reduced oil consumption enough to cause an overall decrease in consumption (Figure 12).

The long-term pattern in which transportation captures an increasingly large share of the petroleum products market continues unabated in the United States and other industrialized nations (Figures 1 and 2). The United States and these other nations can therefore expect the inflexibility of transportation to be an even greater problem during the next gap. Further, because the United States has been relatively successful in finding substitutes for oil in nontransportation sectors, it uses a far larger share of its petroleum products for transportation than do its industrial competitors (Figure 2). Thus the United States has been, and will continue to be, less able to quickly reduce its oil consumption in response to a price shock than have, and will, its industrialized competitors.

C. Summary of Basic Arguments

The fundamental arguments of this paper can be summarized as follows:

- Because of its inability to economically and rapidly substitute nonpetroleum fuels (Figure 6), transportation historically has been less able to reduce petroleum consumption than have other sectors of the industrialized economies (Figures 1 and 2).
- Because of transportation's fuel inflexibility, industrialized nations that devoted a greater share of their total petroleum consumption to transportation had greater difficulty reducing oil consumption after the 1978–1981 crude oil price run-up (Table 1).
- The United States devotes a far larger share of its petroleum consumption to transportation than do its industrial competitors (Figure 2).
- The United States is by far the most dominant transportation oil consumer among OECD nations, using nearly 60 percent of all OECD consumption (Figure 3).
- The long- and short-term evidence shows that the trend in the U.S. is toward use of a greater share of petroleum products by transportation (Figure 1).
- The short-term evidence shows that other industrialized OECD nations are also devoting increasing shares of their petroleum consumption to transportation (Figure 2).

It is, therefore, clear that when the next oil consumption-excess growth gap appears likely, the success or failure of U.S. efforts to reduce consumption by transportation will be more important than ever before in preventing and/or reversing the gap. Because the success of such efforts in other sectors has been greatly enhanced by those sectors' ability to substitute nonpetroleum-based fuels, logic suggests that this capability must be developed in transportation if even more calamitous economic consequences than those in the early 1980s are to be avoided.

VIII. NEED FOR GRADUAL INTRODUCTION OF A PETROLEUM SUBSTITUTE

The problem with developing a substitute fuel for transportation is the high cost and great amount of time needed. If it were not for this high cost, the price run-ups of the 1970s would undoubtedly have brought on the widespread introduction of substitute fuels in transportation. Because many years are needed to smoothly implement new and more costly transportation fuel technologies (Santini, 1989a) — while crude oil price run-ups occur in only a few years or even a single year — the free market cannot respond "efficiently" to crude oil price signals. Consequently, enlightened nonmarket intervention in the transportation fuels marketplace will be necessary if yet another severe price run-up and recession are to be avoided. The key word here is "enlightened," because intervention in the fuels marketplace is the rule rather than the exception. The influence of environmental and safety legislation on fuels markets is as old as the markets themselves. Earlier examples, however, more often contributed to the creation of gaps than to their elimination. (See Santini [1987b] for a detailed discussion of regulatory influence on fuel markets.)

Relatively rapid fuel substitution in U.S. transportation has occurred several times, but never within a period of only a few years. Typically, substitution accelerates in a time of unusually depressed business activity. Major transportation fuel shifts occurred in the late 1830s and early 1840s, the 1870s, the 1890s, and the 1920–1940 period (Santini, 1987b, 1989a). A U.S. depression of several years occurred during each of these periods (Santini, 1987b, 1989a). If the transition from leaded to unleaded gasoline was not major, then no major shift occurred in 1970–1985, but as we have seen, the need for a major shift becomes ever more pronounced as transportation increases its share of petroleum products use and as the United States, and ultimately the world, run out of economically recoverable oil.

A. Methanol: A Possible Mass-Market Substitute Fuel

As the world runs out of oil, both long-term and recent history suggest that natural gas will be the most common source of substitute transportation fuels (Sobey, 1989; Gray and Alson, 1989; Marchetti, 1987). From 1973 to 1987, gas reserves in noncommunist countries increased by 67 percent, while reserves of free world oil increased by only 9 percent

(*Energy Statistics Sourcebook*, 1987). Long-term trends show gas capturing an ever-increasing share of the fuels market. Natural gas was originally found in unwanted association with crude oil and still is often flared at the well or reinjected into the field because of the lack of a market. Because crude oil relatively near the surface has been depleted, many deeper wells drilled to find oil have found gas instead. As mentioned, at greater depths, gas is more likely to be found than crude oil (Mossavar–Rahmani, 1986). Thus, as oil reserves are depleted, a greater proportion of gas is found. Further, recent high prices for natural gas led to more searches for gas for its own sake.

Ironically, the development of an ability to market natural gas as a transportation fuel should increase the amount of oil supplied. Since gas is found with oil in many of the world's reservoirs, something has to be done with the gas as the oil is extracted. Choices are reinjection into the field, flaring (burning the gas as it breaches the surface), venting the gas into the atmosphere, or marketing the gas.

At the present time the technically feasible, demonstrated options for marketing the gas in the transportation sector include use of compressed natural gas (CNG), or chemical conversion of the gas to a liquid. The three liquids presently being used are methanol, methyl tertiary butyl ether (MTBE), or gasoline. Methanol is currently used in Indy racing cars, largely for safety reasons, although it also increases performance for a given engine displacement. It is a widely used chemical feedstock for production of many products and is generally produced from natural gas, but can also be produced from oil, coal, or biomass (plants). MTBE is one chemical which is produced by further processing of methanol, while gasoline can be produced with even further processing of methanol (Swedish Motor Fuel Technology Co., 1986; Sperling, 1988; Sperling, 1989). Gasoline produced from natural-gas based methanol is now produced in New Zealand, so its costs and chemical conversion efficiencies are well known (Swedish Motor Fuel Technology Co., 1986). MTBE is now used as an octane-enhancing additive for U.S. gasolines (primarily the more costly premium grades) and is being considered for use in "reformulated gasoline" in much greater percentages and is expected to see increasing use worldwide, as lead is phased out of gasoline in other nations (*Alcohol Week,* 1990a). Methanol, combined with cosolvents, has also been blended into gasoline, but MTBE, though more costly, has better properties as a gasoline blending component.

The concept of reformulated gasoline has been developed by the auto industry and petroleum industry in part as a method of extending the length of time that the United States can continue to use gasoline in a manner acceptable under environmental regulations. The recently passed Clean Air Act is expected to promote the development and use of reformulated gasolines (*Alcohol Week,* 1990b). For a number of reasons, not the least of which is a likely decline in vehicle sales when new fuels are introduced (I have documented this [Santini, 1989a] and offered a theoretical explanation for the phenomenon in [Santini, 1988a]), the auto industry has an incentive to avoid a switch to fuels that the consumer will perceive to be significantly different from gasoline.

Within the last few years a considerable concern over global warming has emerged. As far as the transportation options being examined here are concerned, the key concern is the total emissions of carbon dioxide (the dominant product of combustion) and methane (natural gas). These two gaseous chemicals warm the atmosphere significantly. Colleagues and I have examined the overall emissions of these two gases under different natural gas originated pathways to the atmosphere (Santini et al., 1989). For vehicles designed to accomplish the same functions as typical U.S. passenger cars, we estimated that the order of global warming damage from pathways putting the carbon in natural gas into the atmosphere is, from worst to best: (1) venting, (2) flaring, (3) gasoline from natural gas, (4) CNG, and (5) methanol. In vehicles designed to have a considerably shorter driving range and lesser acceleration than the typical U.S. passenger car the order of CNG and methanol is reversed. Although MTBE has not yet been analyzed, the facts that it, like gasoline, involves further processing of methanol, while its use will be in gasoline vehicles, imply that its global warming effect will be greater than for methanol.

The reason that gasoline from natural gas is worse than methanol from natural gas is that the total amount of energy used per mile driven is greater, leading to higher cumulative carbon dioxide emissions. The gasoline pathway uses more energy for two reasons. The first is that the methanol must undergo further processing. The second is that gasoline is normally (except when the engine is cold) burned less thermodynamically efficiently than is methanol in an engine designed to take advantage of the properties of the fuel. Thus, in the thermodynamic sense, one loses twice by converting methanol to gasoline rather than burning it in a vehicle designed for methanol. This is also true for MTBE. Thus, if thermodynamic efficiency and global warming were the only issue, methanol fuels would be the preferred way to introduce natural gas into typical U.S. passenger cars.

However, economic efficiency must be considered. In this regard there are significant problems having to do with incompatibility of the existing infrastructure and vehicle stock with methanol fuels, as well as the need to spend more per vehicle mile of service provided on methanol infrastructure than on gasoline infrastructure. The largest part of this difference is due to the fact that methanol has about half the energy per gallon as in gasoline, leading to a need to ship, store, and pump about twice the volume of methanol as gasoline to accomplish the same end (U.S. Department of Energy, 1990). The greater efficiency of combustion of methanol—already demonstrated to be as high as 19 percent—does not completely offset this drawback (Sinor, 1990). Still another drawback is the relatively greater corrosiveness of methanol relative to gasoline and crude oil, making it necessary in most cases to install new pipelines, storage tanks, and refueling pumps to reliably use methanol. Underground tanks must be made of carbon steel or especially formulated fiberglass (U.S. Department of Energy, 1990). The incremental costs (per gallon) of infrastructure compatible with methanol are small when new methanol compatible infrastructure is compared to new gasoline/oil compatible infrastructure. Obviously, however, it would be costly to prematurely replace the existing gasoline-based infrastructure. Such considerations contribute to a desire to reformulate gasoline using MTBE rather than switch to methanol. They also suggest the possibility of an energy/economic security/competition related regulation to require methanol compatibility when gasoline based infrastructure is replaced on a normal schedule.

It has recently been stated that the output of oil from the North Slope of Alaska (*Oil & Gas Journal,* 1990a) and from Saudi Arabia (*Oil & Gas Journal,* 1990b) can be expanded by developing a greater ability to process and use gas. In Saudi Arabia's case the alternative to gas processing was stated to be flaring, which causes far more global warming than use of the gas as a transportation fuel. In the North Sea, some Norwegian oil exploration lease plans were delayed in 1990 because the risk of finding gas instead of oil was judged to be too high (*Oil & Gas Journal,* 1990c). In each of these locations the absence of marketability for the natural gas can either lead to oil being left in the ground or, if the oil is marketed, to dumping the gas into the environment without accomplishing an economically valuable service. Under these circumstances the gas actually has a negative economic value, since more oil would be produced and/or less environmental damage would exist if the gas were not associated with the oil. Economists ignoring these realities are the ones generating the highest estimates of the economically appropriate prices of introduction of natural gas based methanol (Walls and Krupnick, 1990; Schumacher, 1989).

Assuming a positive cost for natural gas feedstocks, a per gallon dealer mark-ups equivalent to premium gasoline, and premium gasoline as the blending stock, SRI International estimated that large scale methanol marketing could result in energy tax equivalent retail sales prices from 83¢ to 98¢ per gallon for M85 fuel (1988$) from Trinidad (Schumacher, 1989). "M85" designates 85 percent methanol and 15 percent hydrocarbons (gasoline-like fuels). Adjusting for the energy content of the fuel and the likely thermodynamic efficiency of the vehicles using it, they estimate that the "gasoline equivalent" retail price would probably be $1.39 to $1.65 per gallon in $1988.

As I have logically argued, in some circumstances natural gas associated with oil should be assigned a negative cost. It is not necessary to blend premium gasoline with methanol in M85. The dealer mark-up that SRI assumed for M85, expressed on a per mile of service provided basis, would be about twice the mark-up on premium gasoline and about four times the mark-up on regular. The average 1989 port of Houston methanol price was 40¢ per gallon (Information Resources Inc., 1989), while SRI's 2007 projections amounted to about 49¢ to 58¢/gallon in 1989$ for the port of Los Angeles (Schumacher, 1989). SRI assumed that a 2007 methanol fuelable vehicle was about 4 percent more efficient than a gasoline vehicle. Nissan's prototype methanol powered Sentra has been estimated by to have 19 percent more efficiency due to use of methanol rather than gasoline under certification test conditions (Sinor, 1990). In real world driving they expect the advantage to be somewhat less. The SRI study was funded by "oil companies marketing in California" (Schumacher, 1989).

This discussion gives a general idea of the gasoline equivalent prices at which methanol fuels could be retailed. These retail prices are far lower than paid in many countries in the world. Based on the points that I made about the SRI numbers one can see that the long-run economic viability of methanol from offshore natural gas is now as much a matter of interpretation of appropriate cost values as of technical uncertainty. It is important to note that the alternative uses of domestic natural gas marketed through existing pipelines generally make domestic natural gas too valuable to turn into methanol as a transportation fuel. Only offshore sources with no local markets can be priced low enough to compete with gasoline (Lawrence and Kapler, 1989).

If methanol cannot be economically competitive unless derived from low (or negative) cost offshore natural gas sources, the question arises, why not use CNG? Experience in New Zealand and Canada with subsidy programs designed to introduce CNG passenger cars shows that even with a 5–10 percent vehicle subsidy and a cost per unit of CNG energy half that for gasoline, CNG vehicles were able to make only limited inroads into the market (Acurex Corporation, 1990; Sauve, 1989; Sathaye, Atkinson, and Meyers, 1989). In Brazil, where ethyl alcohol (ethanol) was substituted for gasoline, prices of ethanol equal per unit of energy to those of gasoline, combined with a 5–10 percent vehicle subsidy led to almost complete substitution of ethanol passenger cars in the new car market in a matter of seven years (Acurex Corporation, 1990; Sathaye, Atkinson, and Meyers, 1989; Trindade and de Carvalho, 1989). This implies that CNG fuel prices per unit of energy need to be about half those of an alcohol fuel in order for CNG to be competitive with the alcohol fuel. Research on the cost of delivering CNG and methanol from offshore locations to the U.S. indicates that they can be delivered for *about the same* cost per unit of energy content (Lawrence and Kapler, 1989). Accordingly, it is doubtful that imported CNG can be competitive with imported methanol, much less oil, in the passenger car market.

Still another reason that imported CNG cannot compete with imported methanol is the absence of a good transitional technology, where the transitional technology makes some compromises to the ideal in order to allow consumers to use either fuel. Both CNG and methanol can be burned more efficiently in a vehicle optimized to use only that fuel. However, to address the problem of initial limitations on the availability of refueling outlets, engineers have developed vehicles capable of burning either gasoline or the alternative fuel. In the case of the natural gas vehicle which can do this, there is a substantial loss of performance when running on natural gas because the engine loses power and a second, heavy refueling system must be added (Sperling, 1989; Santini et al., 1989). In the case of the methanol vehicle with this capability there is a slight increase (about 4 percent) in engine power and thermodynamic efficiency when running on methanol and, when the right materials are used, only one refueling system is needed (Bechtold, Miller, and Hyde,

1990; Sugihara et al., 1990). The U.S. Environmental Protection Agency has already certified one such General Motors vehicle (a Lumina) capable of running on gasoline, methanol, or any combination of the two (*Inside EPA*, 1990). This is termed a flexible fuel vehicle (FFV). In principle the vehicle should also be capable of running on ethanol. Electronic sensors keyed to the chemical composition of the fuel flowing through the fuel lines adjust the timing and other engine parameters to suit the fuel composition being delivered.

In the event that conditions favorable to the introduction of methanol, initially using FFVs, occur in the 1990s, then there may be enough refueling stations in some areas of the country in the following decade to allow sales of vehicles that only burn methanol, or as in Chrysler's gasoline tolerant methanol vehicle, use gasoline only when it is the highest octane premium (Burns, 1989). In that case the full thermodynamic efficiency potential of methanol could be utilized. The fact that Chrysler has designed a methanol vehicle which "tolerates" the highest quality premium hints at the fact that methanol is a potential economic threat to premium gasolines. Ethyl alcohol in Brazil replaced premium gasoline (Sinor, 1990), and Sobey has argued that would be the logical sequence for methyl alcohol in the United States as well (Sobey, 1989).

As any economist who understands competition realizes, this threat affects the oil industry's attitude toward methanol as a fuel. To illustrate, SRI's oil industry funded study assumed costs of methanol as if it were a premium or even super premium fuel and then compared the resulting costs to regular gasoline (Schumacher, 1989). Similarly the oil and refining industries have used advertising to selectively inform the public of the drawbacks of methanol, without illustrating potential benefits.

A key drawback is that methanol is highly poisonous primarily to life forms that can think—primates. It can cause blindness and death in some persons in relatively small quantities. It is odorless, and tasteless and burns with an almost invisible flame (Machiele, 1989; Machiele, 1987; Bauman, 1988; D'Eliscu, 1980; Bevilaqua et al., 1980; Roe, 1982; McMartin, 1988; Litovitz, 1988; Hagey, 1978). Work is ongoing on additives to affect taste, odor, and flame luminosity. Because it burns with a very cool flame and has a much smaller "kill radius" it has been selected for safety purposes at the Indianapolis 500. Ironically, methyl alcohol is more poisonous than ethyl alcohol to primates, but the level of toxicity is reversed for other life forms (Swedish Motor Fuel Technology,

1986; D'Eliscu, 1980; Bevilaqua et al., 1980; Roe, 1982). Methanol is found more commonly in nature than ethyl alcohol, gasoline, or crude oil and most life forms have developed a greater degree of tolerance to it. Humans also have a degree of tolerance and can break down limited quantities of methanol in the body. However, the amount that can be tolerated is far less than for ethyl alcohol. Though the effects of a given quantity on humans can be highly variable, "toxic effects can begin with as little as ... 3 to 4 teaspoons" (Machiele, 1989). Unfortunately it tastes like ethyl alcohol and has the same initial effect, so it must be denatured or clearly identified to avoid accidental ingestion. When spilled it is generally much less damaging and much less permanent than spilled crude oil or petroleum products (Swedish Motor Fuel Technology, 1986; D'Eliscu, 1980; Bevilaqua et al., 1980).

When used in a vehicle, methanol and modified catalysts offer the potential to reduce emissions of pollutants such as carbon monoxide, nitrogen oxide, and hydrocarbons. The potential to reduce hydrocarbon emissions involves both evaporative emissions and products of combustion (Alson, Adler, and Baines, 1989). The potential to reduce "reactive" hydrocarbons and nitrogen oxides, both precursors of ozone, has been one reason for strong interest in methanol by the U.S. EPA and the California state government. The chemistry involved is complex and there is debate over the degree of advantage potentially offered for methanol (Russel, 1989). While improvements in average ozone concentrations in Los Angeles is likely if methanol fueled vehicles are introduced, it has been argued that conditions in the Houston area would worsen. From the perspective of energy cost and/or energy security, however, a vehicle as good (on average) as a gasoline vehicle is acceptable. A drawback of methanol is worse emissions when the engine is cold and generally greater emissions of aldehydes than for gasoline. New catalyst designs are being worked on to minimize this problem. In cold climates methanol engines are more difficult to start than gasoline engines. Consequently, initial introduction is more logical in warmer climates. However, with an FFV, it would be possible to fuel with higher percentages of gasoline in winter months. Dash mounted sensors will inform drivers of the percentages of gasoline and alcohol in the fuel line.

This discussion shows that the potential introduction of methanol involves very complex economic, environmental and safety trade-offs. Familiarity with those trade-offs causes me to reach the conclusion that selective introduction of methanol into certain U.S. markets in the 1990s would be a sound economic and environmental step for the United States to take. The resulting threat of such introduction to the worldwide oil industry in the next century is not that methanol will (or should) soon replace oil and gasoline as the dominant fuel. Rather the threat is that competition will be introduced, reducing profits by increasing the options available to consumers and the base of supply from which they may draw their transportation fuels.

IX. CLOSING

Since the article contributing most of the substance of this paper was completed (Santini, 1988b), conditions have changed significantly and the knowledge base improved substantially. Global warming has become an issue driving the discussion of long-term fuel choices. Where methanol from natural gas was previously suggested as a transition option to methanol from domestic coal (Gray and Alson, 1989), the high global warming effects of coal-based transportation fuel pathways is pushing this proposed transition out of consideration. Natural gas-based methanol is now being considered as a transition fuel to biomass-based methanol (Takashi, 1990), which is estimated to have no net global warming effect, and also allow eventual substitution of domestic biomass-based fuels for imported fuels. Our ongoing analyses of the trade-offs suggest that converting the gasoline based infrastructure to compatibility with natural-gas based methanol would be a far better alternative from the point of view of global warming than would an attempt to keep the gasoline-based infrastructure and "force" compatible natural gas-based liquids (gasoline, MTBE) into that infrastructure (Santini et al., 1989).

From the short-term economic perspective, the arguments of this paper have been strengthened by the fact that the gap, as defined here, widened to historically extreme values in the last years (Figure 4) and a sharp oil price increase followed. The earlier article warned of the implications of this pattern, which was emerging at the time. In a subsequent conference paper I anticipated the implied risk of a price increase and the negative consequences for the economy (Santini, 1989b). Thus, the gap seems to have passed its second test as a predictive tool. The first was in 1985–1986 when I predicted the oil price collapse and the subsequent weak macroeconomic behavior, using the gap concept then under development (Santini, 1985).

Although significant research progress has been made in the area of evaluating and demonstrating alternative transportation fuels to petroleum, and many forceful informed advocates are being heard, recent government decisions lead me to have doubts that alternative transportation fuels will be significantly pushed into the market in any other way than by another expansion of the gap, an oil price run-up, and attendant macroeconomic difficulties. Nevertheless, it is too early to declare failure. It should be emphasized that the tampering with the market should be the minimum necessary and the analysis supporting any intervention should be thorough. While I advocate the early introduction of alternative fuels to a greater extent than now appears likely, I also caution against excessively aggressive or misdirected programs as well (Rajan and Santini, 1990). Brazil "successfully" replaced gasoline fuels with alcohol fuels in nearly 100 percent of the new passenger cars sold in a matter of seven years. However, Brazil now recognizes that this effort was too aggressive and is now suffering from high costs imposed by a too rapid alteration of the refining industry to cope with a huge change in the mix of diesel fuel and gasoline required by consumers (Borges, 1990). Further, my recent research implies that the complete substitution of fuels within the dominant vehicles produced by the United States has consistently retarded economic growth during the transition (Santini, 1988a, 1989a). This has clearly happened in Brazil as well, as average growth rates for vehicle sales (Sathaye, Atkinson, and Meyers, 1989; Trindade and de Carvalho, 1989), manufacturing output, and GNP (World Bank, 1990) dropped to a far greater extent in Brazil than in the vast majority of nations during the transition period (1980–1988 growth compared to 1965–1980). As I have stated elsewhere (Santini, 1989a), it is my judgment that alternative fuels programs should start earlier and proceed more slowly than would be the case if oil price signals are to be the guiding force. If I am correct, the gap can be used as a surrogate early indicator of the need to shift national resources. Appropriate use of and response to this indicator should reduce the magnitude of energy price swings and the depth of the worst troughs in the business cycle.

ACKNOWLEDGMENT

This paper is authored by a contractor of the U.S. government under contract No. W-31-109-ENG-38. The work draws from research sponsored by three offices at the U.S. Department of Energy: the Office of the Environment, the

Office of Policy, Planning and Analysis, and the Office of Transportation Technologies. Much of this paper is a nearly exact reproduction of Santini (1988b). Permission to republish this material has been granted by the Transportation Research Board, National Research Council, Washington, DC. The opinions expressed here are those of the author and do not necessarily represent the positions of the U.S. Department of Energy, Argonne National Laboratory, nor the Transportation Research Board.

REFERENCES

Acurex Corporation, Environmental Systems Division. 1990. *Mandates and Incentives Report*, Vol. 5.

Alcohol Week. 1990a. "Latin America, Caribbean Countries Expected to be Major MTBE Exporters to U.S." *Alcohol Week* 11, 44(November 5): 9–10.

Alcohol Week. 1990b. "Alternative Fuels Seen Greatly Buoyed Due to Approval of Major Legislation." *Alcohol Week*, 11, 44(November 5): 5.

Alson, J.A., J.M. Adler, and T.M. Baines. 1989. "Motor Vehicle Emission Characteristics and Air Quality Impacts of Methanol and Compressed Natural Gas." In *Alternative Transportation Fuels: An Environmental and Energy Solution*, edited by D. Spering. Westport, CT: Quorum Books.

Amdall, J.K. "Initial Operating Results from Caterpillar 3600 Diesel Engine", American Society of Mechanical Engineers Paper 87–ICE–30, presented at the Energy-Sources Technology Conference and Exhibition, Dallas, February 15–20.

American Petroleum Institute. *Basic Petroleum Data Book: Petroleum Industry Statistics* (various issues). Washington, DC: American Petroleum Institute.

Bauman, B. 1988. "Groundwater Contamination by Methanol Fuels." Pp. 21–26 In *Proceedings of the Methanol Health and Safety Workshop, South Coast Air Quality Management District, Los Angeles*. San Mateo, CA: Polydyne Inc.

Bechtold, R.L., M.T. Miller, and J.D. Hyde. 1990. *Ford Methanol FFV Performance/Emissions Experience*. SAE Paper 902157.

Bevilaqua, O.M. et al. 1980. *An Environmental Assessment of the Use of Alcohol Fuels in Highway Vehicles*. Argonne National Laboratory Report No. ANL/CNSV–14, Argonne, IL: Argonne National Laboratory.

Borges, J.M.M. 1990. "The Brazilian Alcohol Program: Foundations, Results and Perspectives." *Energy Sources* 12: 451–461.

Burns, V.R. 1989. "Chrysler's Flexible-Fuel and Gasoline Tolerant Methanol Vehicle Development," Paper presented at the 1989 SAE Government/Industry Meeting, Washington, DC, May 2–4.

D'Eliscu, P.N. 1980. "Environmental Consequences of Methanol Spills and Methanol Fuel Emissions on Terrestrial and Freshwater Organisms." In *Proceedings: Third International LIB Symposium on Alcohol Fuels Technology*. U.S. Department CONF–790520. Washington, DC.

Dolan, E.G. 1986. *Macroeconomics*, 4th ed. Chicago: Dryden.

Dunkerley, J., and I. Hoch. 1986. *The Pricing of Transport Fuels, Energy Policy*: 307–317.

Dunstan, A.E. et al. 1938. *The Science of Petroleum*. London: Oxford University Press.
Energy Statistics Sourcebook. 1987. Tulsa, OK: PennWell Publishing.
Gray, C.L. and J.A. Alson. 1985. *Moving American to Methanol*. Ann Arbor: University of Michigan Press.
Gray, C.L., and J.A. Alson. 1989. "The Case for Methanol." *Scientific American*, 252, 11(November): 108–114.
Hagey, G. et al. 1978. "Methanol and Ethanol Fuels—Environmental Health and Safety Issues," In *Proceedings: International Symposium on Alcohol Fuel Technology*. U.S. Department of Energy CONF–771175. Washington, DC.
Information Resources Inc. 1989. *Alcohol Update* (all issues). Washington, DC.
Inside EPA. 1990. "EPA Certifies Methanol-Blend Vehicle; Fleet Use Expected In California." *Inside EPA*, 1, 13(October 25): 18.
Investor's Daily. 1987. Based on a statement attributed to OPEC President Rilwanu Lukman (November 18).
Kahn, G.A. 1987. "Dollar Depreciation and Inflation." *Economic Review of the Federal Reserve Bank of Kansas City*, 72, 9(November): 32–49.
Lawrence, M.F., and J.K. Kapler. 1989. "Natural Gas, Methanol and CNG: Projected Supplies and Costs." In *Alternative Transportation Fuels: An Environmental and Energy Solution*, edited by D. Spering. Westport, CT: Quorum Books.
Litovitz, T. 1988. "Acute Exposure to Methanol in Fuels: A Prediction of Ingestion Incidence and Toxicity," In *Proceedings of the Methanol Health and Safety Workshop, Los Angeles*. San Mateo, CA: Polydyne Inc.
Machiele, P.A. 1987. "Flammability and Toxicity Tradeoffs with Methanol Fuels" (SAE Technical Paper No. 862064). Paper presented at the International Fuels and Lubricants Meeting and Exposition, Toronto, Ontario.
Machiele, P.A. 1989. "A Perspective on the Flammability, Toxicity, and Environmental Safety Distinctions Between Methanol and Conventional Fuels." Paper presented at the American Institute of Chemical Engineers 1989 National Summer Meeting, Philadelphia, August 22.
Marchetti, C. 1987. "The Future of Natural Gas: A Darwinian Analysis." *Technological Forecasting and Social Change*, 31, 15–171.
McConnell, C.R. 1987. *Economics: Principles, Problems, and Policies*, 10th ed. New York: McGraw-Hill.
McMartin, K.E. 1988. "Metabolism, Ocular Toxicity, and Possible Chronic Effects of Methanol." In *Proceedings of the Methanol Health and Safety Workshop, Los Angeles*, San Mateo, CA: Polydyne, Inc.
McNutt, B.D., and E.E. Ecklund. 1986. "Is There a Government Role in Methanol Market Development?" (Society of Automotive Engineers Technical Paper 861571). Paper presented at the International Fuels and Lubricants Meeting and Exposition, Philadelphia, October 6–9.
Merrill, W.C., and K.A. Fox. 1970. *Introduction to Economic Statistics*. New York: Wiley.
Mossavar–Rahmani, B., and S. Mossavar–Rahmani. 1986. *The OPEC Natural Gas Dilemma*. Boulder, CO: Westview.
Oil & Gas Journal. 1990a. "IPAA: 1991 Destined to be Sixth Year of Declining U.S. Oil Production." *Oil & Gas Journal*, 88, 44(October 29): 19.

Oil & Gas Journal. 1990b. "Oil Prices Plunge as Persian Gulf Crisis Eases." *Oil & Gas Journal,* 88, 44(October 29): 19.

Oil & Gas Journal. 1990c. "Norsk Hydro Taps Oil Pay in Troll Gas Field." *Oil & Gas Journal,* 88, 19(May 14): 34.

Organization for Economic Co-operation and Development. 1984. *Energy Balance of OECD Countries, 1971–81.* Paris: OECD.

Rajan, J.B., and D.J. Santini. 1990. "Comparison of Emissions of Transi Buses Using Methanol and Diesel Fuels" (Paper No. 890396). Paper presented at the Transportation Research Board 69th Annual Meeting, Washington, DC, January 7–11.

Roe, O. 1982. "Species Differences in Methanol Poisoning." *CRC Critical Review in Toxicology,* 10, 4.

Rossi, C.E. 1986. "Research in the Automotive Industry." *Proceedings of the Institution of Mechanical Engineers,* 200, D2: 149–158.

Russel, A. 1989. "Air Quality and Alternative Fuels." Paper presented at the American Institute of Chemical Engineers 1989 National Summer Meeting, Philadelphia, August 22.

Santini, D.J. 1985. "Unbalanced Recovery in Energy Related Markets: Growth Retarding Implications for 1985." Paper presented at the Fifth International Symposium on Forecasting, Montreal June 9–12.

Santini, D.J. 1986. *The Petroleum Problem: Managing the Gap.* Argonne National Laboratory Report ANL/CNSV–58, Vol. 2 (August). Argonne, IL: Argonne National Laboratory.

Santini, D.J. 1987a. *Verification of Energy's Role as a Determinant of U.S. Economic Activity.* Formal Report ANL/ES–53 (October). Argonne, IL: Argonne National Laboratory.

Santini, D.J. 1987b. "Micro- and Macroeconomic Responses to Energy Price Shocks: A Discussion of Past, Present, and Future Economic Problems in Simultaneously Introducing Alternative Fuels and Major Efficiency-Enhancing Vehicular Engine Innovations", Paper presented at the Operations Research Society of America and Institute of Management Sciences Joint National Meeting, St. Louis, MO, October 22–28.

Santini, D.J. 1988a. "A Model of Economic Fluctuations and Growth Arising from the Reshaping of Transportation Technology." *The Logistics and Transportation Review,* 24, 2: 121–151.

Santini, D.J. 1988b. "The Past and Future of the Petroleum Problem: The Increasing Need to Develop Alternative Transportation Fuels." *Transportation Research Record,* 1175:1–14.

Santini, D.J. 1989a. "Interactions Among Transportation Fuel Substitution, Vehicle Quantity Growth, and National Economic Growth." *Transportation Research,* 23, 3 (Part A): 183–207.

Santini, D.J. 1989b. "The Slowdown of 1989–90: "Further Evidence for the Use of Transportation Sector Indicators to Predict GNP Growth." Paper Presented at the Fourth Annual Conference on Business Forecasting, September 27–28.

Santini, D.J. et al. 1989. "Greenhouse Gas Emissions from Selected Alternative Transportation Fuels Market Niches." Paper presented at the American Institute of Chemical Engineers 1989 National Summer Meeting, Philadelphia, August 22.

Sathaye, J., B. Atkinson, and S. Meyers. 1989. "Promoting Alternative Transportation

Fuels: the Role of Government in New Zealand, Brazil and Canada." In *Alternative Transportation Fuels: An Environmental and Energy Solution*, edited by D. Spering. Westport, CT: Quorum Books.

Sauve, R. 1989. "Compressed Natural Gas and Propane in the Canadian Transportation Energy Market." In *Alternative Transportation Fuels: An Environmental and Energy Solution*, edited by D. Spering. Westport, CT: Quorum Books.

Schumacher, W.J. 1989. *The Economics of Alternative Fuels and Conventional Fuels.* Menlo Park, CA: SRI International.

Sekar, R., and L. Tozzi. 1983. *Advanced Automotive Diesel Assessment Program.* NASA–Lewis Report DOE/NASA/0261–1 NASA CR–168285 CTR 0747–84003. Cleveland, OH: NASA Lewis Research Center.

Sinor, J.E., Inc. 1990. *The Clean Fuels Report.* Vol. 2, No. 3. Niwot, CO: J.E. Sinor Consultants Inc.

Sobey, A.J. 1989. "A Global Fuels Strategy: An Automotive Industry Perspective." Pp. 205–219 in *Alternative Transportation Fuels: An Environmental and Energy Solution*, edited by D. Sperling. Westport, CT: Quorum Books.

Society of Automotive Engineers, Inc. 1990. *Methanol Fuel Formulations and In-Use Experiences.* SAE Special Publication SP–840.Warrendale, PA: Society of Automotive Engineers.

Sperling, D. 1988. *New Transportation Fuels: A Strategic Approach to Technological Change.* Berkeley: University of California Press.

Sperling, D., ed. 1989. *Alternative Transportation Fuels: An Environmental and Energy Solution.* Westport, CT: Quorum Books.

Sugihara, K. et al. 1990. *Research and Development of Flexible Fuel Vehicles at Nissan.* SAE Paper 902159.

Swedish Motor Fuel Technology Co. 1986. *Alcohols and Alcohol Blends as Motor Fuels.* Vols. I, IIA, and IIB. Paris: International Energy Agency.

Takashi, P.K. et al. 1990. "Hawaii: An International Model for Methanol From Biomass." *Energy Sources*, 12: 421–428.

Trindade, S., and A.V. de Carvalho, Jr. 1989. "Transportation Fuels Policy Issues and Options: The Case of Ethanol Fuels in Brazil." In *Alternative Transportation Fuels: An Environmental and Energy Solution*, edited by D. Spering. Westport, CT: Quorum Books.

U.S. Bureau of the Census. *Statistical Abstract of the United States.* Washington, DC: U.S. Government Printing Office.

U.S. Department of Commerce. 1975. *Historical Statistics, Colonial Times to 1970.* Washington, DC: U.S. Government Printing Office.

U.S. Department of Commerce. Bureau of Economic Analysis. 1987. *Survey of Current Business.* Washington, DC: U.S. Government Printing Office.

U.S. Deptartment of Energy. Energy Information Administration. *Monthly Energy Review* (various issues). Washington, DC: U.S. Government Printing Office.

U.S. Department of Energy. 1990. *Assessment of Costs and Benefits of Flexible and Alternative Fuel Use in the U.S. Transportation Sector.* Technical Report 4: *Vehicle and Fuel Distribution Requirements.* Washington, DC: U.S. Department of Energy, Office of Policy, Planning, and Analysis.

U.S. Department of Transportation. Federal Highway Administration. 1986. *Highway Statistics: Summary to 1985.* Washington, DC: U.S. Government Printing Office.

Walls, M.A., and A.J. Krupnick. 1990. "Cost Effectiveness of Methanol Vehicles." *Resources for the Future*, 100 (Summer): 1–5.

Williamson, H.F., and A.R. Daum. 1959. *The American Petroleum Industry: The Age of Energy, 1899–1959*. Evanston, IL: Northwestern University Press.

World Bank. 1990. *World Development Report 1990: Poverty*. Oxford, England: Oxford University Press.

ENERGY AND THE MACROECONOMY: CAPITAL SPENDING AFTER AN ENERGY COST SHOCK

D. J. Santini

ABSTRACT

A two region, three period model of the process of capital-energy substitution in response to an energy price shock is constructed. The model predicts that statistical analyses done in cross-section will correctly estimate that energy and capital are substitutes, while those done in time series will incorrectly estimate that capital and energy are complements. The model uses a derivative putty-clay formulation to show how thermodynamic-efficiency-enhancing capital-for-energy substituting innovation will occur after an energy price shock in conjunction with a temporary decline in capital spending. The greater the aggregate increase in thermodynamic efficiency of the capital stock, the greater the predicted decline and duration of depressed capital spending.

I. INTRODUCTION: ENERGY'S EFFECT ON THE MACROECONOMY

A. Short Run (Year to Year) Evidence—Recession Cases

Energy's clear role in the macroeconomic difficulties of the 1970s has prompted a reconsideration of the role that energy plays in both micro- and macroeconomic behavior. As early as 1981, Tatom presented estimates and a simple model of the relationship between energy prices and short-run economic performance. Separately and jointly estimating Granger causality tests for the periods 1948–1972 and 1973–1980, Hamilton (1983) has found evidence that crude oil price shocks are statistically significant predictors of declines in GNP and increases in unemployment, with a four quarter lag. Hamilton's univariate test results, which are based on post-World War II data, also indicate that crude oil price shocks are better predictors of shifts in several key macroeconomic variables than are the generally used predictors. Loungani (1986a,b) found that oil, lagged one year (e.g., four quarters) remained a statistically significant predictor of unemployment when tested in a multivariate regression with money for the 1948–1980 period. Loungani also found that real energy prices were significantly related to unemployment in a money plus energy regression estimated for the 1901–1929 period, but found no relationship during the Great Depression (1986a).

Santini (1986a, 1987a) using Granger causality tests with annual data, separately confirmed Hamilton and Loungani's results for GNP and unemployment for the post-World War II period, but, like Loungani, found that crude oil price shocks were not a useful predictor during the 1929–1940 period. Santini found, however, that a more general energy-based formulation, in which changes in personal consumption expenditures on energy were used as a predictor, worked far more consistently than wellhead and mine-mouth energy prices in six non-wartime subperiods from 1890–1982, including two depressions (Santini and Belak, 1985). Santini (1985a) constructed a preliminary, Keynesian, consumption based model which provides one explanation for Hamilton's, Loungani's and Santini's statistical findings. This model, termed the "energy squeeze" model, relies on the fact that increased consumer costs for energy cause a decline in consumption of other goods. This effect can arise from either an energy price shock (an increase of energy price relative to aggregate prices) or from an unusually severe winter, or both. The price shock can occur at the point of extraction, as Hamilton found, or further upstream in the energy supply system. The upstream energy cost increasing effect, termed a "spiral of impossibility" occurs during severe downturns when declining demand forces capital (and debt) intensive energy industries to raise retail prices (or reduce them by less in a deflationary environment) when demand declines. This effect, and some of the literature describing it are discussed by Santini (1985a). Hamilton's results imply that the macroeconomic side effects of a winter time "energy squeeze" are likely to show up in the following winter.

Even though there is no widely accepted theory relating energy to macroeconomic behavior, consideration of energy's role in the behavior of the economy is not uncommon. Adam Smith (1976) deduced that energy prices were worthy of repeated discussion, even though he did not see them as a fundamental element in the theories that he developed in *The Wealth of Nations*. In light of the above discussion, his statements concerning the importance of winter fuel expenses are noteworthy. He observed that "on account of the extraordinary expense of fuel, the maintenance of a family is most expensive in winter. A laborer, it may be said indeed, ought to save part of his summer wages in order to defray his winter expense." While Smith was consistently concerned with the long run welfare of the macroeconomy, he felt this short run consumer energy budgeting problem deserved discussion.

This problem alone may account for Hamilton's results. Since consumers and weathermen have a limited ability to predict the severity of a coming winter, it is reasonable to expect consumers and businesses to cope with the energy cost involved in getting through a cold winter by borrowing during the winter and reducing future consumption in order to repay those debts. Further, a lagged savings reaction effect might be expected, in which consumers and businesses might save more than before in preparation for the following winter. When combined, these tendencies to repay increased debts and save more against the future could easily cause a decline in economic activity about one year later, as Hamilton (1983), Loungani (1986a,b), Tatom (1981), and Santini (1983a) observed. Though these studies showing important short-run energy to macroeconomy effects are new, the longer run economic importance of energy costs have been recognized by other economists periodically in history.

B. Occasional Long-Run Effects Including Depression Cases

In the 1860s Jevons correctly predicted the relative economic exhaustion of much of Britain's coal resources in a century's time (Jevons, 1965). He also correctly predicted that, to the extent that the British economy continued to be dependent on coal, a decline in the position of the British economy relative to those of other nations was inevitable. His initial prediction was made during the U.S. Civil War, a period of unusually high U.S. coal prices. Jevons' prediction was followed by a run up in British coal prices that occurred from 1869–1973, after completion of the Suez canal. The canal allowed substitution of steamships for sailing ships on the shortened Britain to India route. The resulting increase in coal demand caused a run-up in British coal prices which promoted the successful adoption of the much more efficient compound marine steam engine. The Suez canal and the compound engine were said by Greenhill (1980) to complement one another, each contributing to the economic success of the other. Once this more efficient engine was widely adopted, British coal prices collapsed.

This price collapse was also promoted by widespread adoption, in both the United States and Britain, of the Bessemer process, an extremely efficient method of producing steel. The Bessemer process, described by Bessemer in 1856 as the "Manufacture of Malleable Iron and Steel Without Fuel" (see Rosenburg, 1982), used one seventh the fuel used by the most popular competing method. The 1873 collapse in coal prices occurred in conjunction with a financial panic in the United States and Germany. Germany, the chief exporter of steel rails to the United States, was adversely affected by the new technology because its coals were not compatible with the new process. The success of the British economy from 1873–1875 was aided by the fact that Britain possessed, and built, the most efficient newly introduced technologies.

These technologies made competing U.S. and German technologies obsolete. However, the period following the most rapid rate of introduction of the Bessemer process and the compound marine engine was followed by depressions in the United States (1873–1878) and Britain (1875–1881). The model proposed here reproduces this sequence of events in an abstract way.

Because he did not emphasize the trees while describing the forest, Jevons has recently been incorrectly ridiculed for ignoring the value of price induced substitution of new technologies for the coal intensive technologies of the time (Simon, 1981). Obviously, some of the many such substitutions took place soon after Jevons made his projections. In fact, however, Jevons explicitly stated that efficiency enhancing technologies could be expected, but he used exponential growth and exhaustible resource arguments to convince himself that they would only forestall the inevitable decline of the British economy relative to other economies. He pointed out that the introduction of the "hot blast" iron production technique resulted in one-third the coal use per ton but "was followed, in Scotland, by a tenfold total consumption, between the years 1830 and 1863" (Jevons, 1965, p.154). He stressed the importance of "comparative cheapness of " Britain's "fuel and motive power" (p. 156).

Note that Jevons included *and motive power*. He was alluding to the need to maintain Britain's then superior position in energy conversion as well as energy extraction. This was a challenge he expected to be very difficult because of the ease with which competitors could copy technology. The relatively poor performance of Britain's economy in the twentieth century compared to the nineteenth is not only a function of uneconomic coal-mining, but also a loss of dominance in producing engines and vehicles. When the term and motive power is added to Jevons' arguments, the current superior behavior of Japan's economy is partly explainable within the broad context of Jevon's energy arguments. With half the population, Japan produces about as many automobiles as the United States. The two nations are now competitors for world leadership in vehicle production, with Britain a far distant competitor. When Jevons issued his report Britain was the world's leader in "motive power."

Jevons' carefully couched and correctly stated general findings were

> In the increasing depth and difficulty of coal mining we shall meet that vague but inevitable limit which must stop our progress. So far, then, as our wealth and our progress depend upon the superior command of coal we must not only cease to progress as before—we must begin a retrograde career (Jevons, 1965, pp. 200–201).

His arguments were thus very long run in nature. When Britain's twentieth-century economic stature relative to the United States, Japan, and Germany is considered compared to the nineteenth century, Jevon's arguments have a ring of truth.

Hotelling (1931) wrote what is now recognized as a classic article on depletion of exhaustible resources after several years of concern in the

United States over the possible depletion of crude oil supplies. Like Jevons, Hotelling's findings concerning energy were published within a decade when the energy resource of concern to the nation experienced a price collapse (in the midst of deflation, the price of U.S. crude oil relative to wholesale prices declined by 35 percent from 1930–1931, even though relative gasoline prices increased). Given the energy price collapses following the works of Hotelling and Jevons, the response to these publications concerned with fuel depletion was not overwhelming.

The long-run view allows the interpretation that the concerns over energy resource depletion expressed by Jevons and Hotelling were, if not unwarranted, excessive. However, as Keynes is purported to have pointed out, in the long run we are all dead. Those economists who draw comfort from the fact that the resources that spurred Jevon's and Hotelling's concern proved to be abundantly available in the long run would do well to remember that the energy price collapses in these two cases occurred near the onset of U.S. and British depressions. Even the 1981–1986 declines in oil prices were initially promoted by macroeconomic behavior described by some economists as a depression.

Having observed the general long-run pattern of energy price increases, engine innovation, and subsequent energy price collapse not only in the cases cited here, but in other U.S. depressions as well, I initially proposed a comprehensive, but non-mathematical, theory which argues that economic responses to energy price shocks are a critically important determinant of the business cycle (Santini, 1983a,b,c).

Since that time, having studied the short-run events that accompanied Jevons' and Hotelling's publications, I have been able to avoid some of the pitfalls of arguing that energy is crucially important to economic success at about the time energy problems seem to be solved. I anticipated and predicted the 1985–1986 oil price collapse and near recession that followed (Santini, 1985b). I also anticipated the 1989–1990 oil price run-up and its consequences (Santini, 1989a). The 1985–1986 oil price collapse and the more recent oil price run-up have contributed to regional financial difficulties which continue to plague U.S. financial institutions. Fortunately, the 1985–1986 oil price collapse did not coincide with the earlier auto sales collapse (1978–1982) as happened at the start of the Great Depression (1929–1932).

While the assertion that energy price shocks are a key determinant of the business cycle is new, the general idea that microeconomic adjustments to specific price shifts have more of a causal relationship to the business cycle than do movements of price and money aggregates is not new. Perhaps the most notable proponent of this view, Frederich Hayek, developed his theory during the same years that Keynes was working on *The General Theory* (Hayek 1979). Hayek, however, does not attempt to link his theory to a particular set of price movements. His arguments include a "regional" perspective incorporating a logical incompatibility between national attempts at stabilization during periods of international structural realignment of the means of production. The prior discussion of international effects of introduction of the Bessemer process is a tangible example of the theoretical problem Hayek describes. No monetary policy could have completely compensated for the fact that German coals were incompatible with the Bessemer process.

I have written several non-mathematical papers to document much of the U.S. historical evidence in support of the theory that the microeconomic behavior of the energy-intensive sectors is the cause of the business cycle (Santini, 1983a,b,c, 1984, 1985c,d, 1987b). My evidence concerning the importance of energy prices is confined primarily to the U.S. economy. This U.S. evidence may, therefore, be regarded only as a partial verification of Hayek's more general theory. More recently, I have begun to formalize certain aspects of the theory in limited mathematical models (1985a, 1986b, 1988). However, these models do not separately or jointly constitute a "general" theory. For example, some elements of my empirical work cannot be explained by the "energy squeeze" consumer response model (1985a) previously discussed. This paper, and its more lengthy predecessor (1986b) are intended, in part, to correct that shortcoming. My 1988 paper "A Model of Economic Fluctuations and Growth Arising From the Reshaping of Transportation Technology" offers still another unique model for consideration.

I distinguish depressions and stagnations from recessions in terms of the difficulties in making a transition from one set of energy intensive technologies to another, more efficient set. The prior discussion of technology transitions in the 1870s is illustrative of the kind of observations that I have made about every U.S. depression. However, my historical studies of technology transitions during U.S. business cycles focus primarily on transportation technologies. Prior to World War I the final consumer had very little to do with the selection of technologies used in national and international transportation systems. Consequently, while my "energy squeeze" consumer model is useful as a partial explanation for linkages of energy price shocks and business downturns, it does not successfully explain my pre-World War I empirical findings for the transportation sector.

Another attribute of my general description of the theoretical causal relationship between energy technology transitions and depressions/stagnations is the multiyear length of the transition period. A simple one year/four-quarter consumer reaction model cannot adequately explain a several year period of economic difficulty by itself. However, several years of consistent exogenous pressure on energy prices could provide such an explanation. Several consecutive years of cold winters or unsuccessful energy exploration could theoretically create such an effect. The argument that energy problems are causes of depressions/stagnations would be greatly strengthened if one could logically show that microeconomic decisions by businesses were affected in such a way that investment in capital goods would decline, perhaps for several years, as a result of energy related problems experienced by those businesses. Further, if the explanation for such decisions were derived from standard economic theory, the explanation would be more satisfying and more readily integrable into existing basic theory.

It is the purpose of this paper to present a partial two-region model of firms' response to energy price shocks that is general in nature and is derived from the most basic microeconomic theory of the firm. This model is designed specifically to explain how firm responses to energy price shocks can contribute to recessions. A detailed depression case extension of this model can be found in the original manuscript (Santini, 1986b), from which this paper is derived. A brief discussion of the depression case is given in closing.

II. THE RECESSION CASE MODELLING PROBLEM

Gibbons (1984, 1985) has addressed the question of substitutability of capital for energy, using "restrictive" putty-clay assumptions concerning the nature of the energy intensive capital stock. Gibbons' model predicts that capital should be substitutable for energy. His arguments and his model predict that escalation of energy prices should stimulate investment. His review of empirical studies on the subject, however, shows that the results obtained in cross-section and time series are inconsistent. Specifically, his review shows that, in cross section, regions with higher energy prices tend to employ more capital, resulting in estimates of a positive cross elasticity of demand between price of energy and spending on capital. In time series studies, though, these results are reversed. As Gibbons puts it, time series studies result in the "gloomy finding" that energy and capital are complements. Gibbons' review also shows that pooled time-series cross-section studies give intermediate results. Gibbons trusts his derived theoretical implications rather than the time series results, maintaining that the long term consequences of the 1970s oil price shocks will be substitution of capital for energy. Nevertheless, he offers no structural model (or submodel) to explain why the very short term (i.e., annual and quarterly) empirical time series results are incompatible with the substitutability implications derived from his model. In light of the inconsistent empirical results concerning substitutability of capital and energy, it is not surprising that modelers sometimes assume "zero" substitutability for capital and energy (Mork and Hall, 1980).

The resolution of the time-series versus cross-section inconsistencies is a critical problem. If both results are correct, then a theoretical explanation is necessary for both policy and modelling purposes. If the short-run time-series results mask a long run ability to substitute capital for energy, the knowledge of this long-run capability may be necessary for proper formulation of economic growth models. Gibbons' assumption that the substitutability of capital for energy will show up eventually in the 1980s involves another implicit assumption made by most economists. That assumption is that the energy problems of the 1970s are historically atypical. Such an assumption is made when economists ignore the possibility that a study of historical data can reveal the general nature of the economy's long-term response to energy price shocks.

My historical studies suggest that Gibbons' work is very close to the truth in its present form. My less formal model of economic behavior and his empirical results show that capital is substituted for energy in the *long-run* (over decades) in response to energy price shocks, but that *short-run* (intra decade) responses involve a decline in business and consumer purchases of energy intensive technologies. Lareau and Darmstadter (1982) have shown very clearly that the modern day consumer has a very high short-run (year-to-year) negative cross energy price elasticity of demand for energy intensive commodities, especially automobiles. My long-run findings are consistent with Gibbons' prediction of long-run substitutability, as well as the observed short-run empirical estimates of complementarity.

In this paper, a formal theoretical description of this process is constructed using a model very similar in its starting structure to Gibbons model. The model also involves a fairly straightforward derivation from the theory of the firm.

III. THE BASIC MODEL

In order to examine both cross section and time series situations, the model involves three periods of time, t, and two regions, r. The first and third periods are periods when the economy is in general equilibrium. The intermediate period is a period of disequilibrium induced by an energy price shock. The intermediate period may be thought of as the period of transition from one set of energy intensive technologies to a new, more efficient set. During the two equilibrium periods all firms are assumed to earn normal profits. They just cover their fixed and variable costs, earning region (r) and time-specific (t) revenue per unit output equal to R_{rt}. No economic profits exist. During the disequilibrium period four distinct steps of adjustment occur. In the initial step, old product prices are maintained while an energy price rise temporarily causes unit costs of production to rise above R, such that all firms initially operate at a loss. During this first step of the disequilibrium period each firm earns enough revenue to pay variable energy costs, but does not earn enough to pay fixed capital costs. In the abstract world posited in this model revenue per unit output is allocated only to quantities, q, of utilized capital and energy, according to the following equation:

$$R_{rt} = P_k q_{ki} + P_{ert} q_{ei} \qquad (1)$$

where: P_k = the real price of capital, which is assumed to be constant geographically and temporally;
q_{ki} = the quantity of capital, k, utilized per unit output for technology i ;
P_{ert} = the price of energy, e, in region r at time t; and
q_{ei} = the quantity of energy, e, utilized per unit output for technology i

One of the important differences between this model and the putty-clay model used by Gibbons (1984) and Berndt and Wood (1975, 1979) is the assumption that the cost of capital is constant. The purpose of this assumption is to study the behavior of the firm under conditions where the price of energy is far more variable than the price of capital (both geographically and temporally). Gibbons predicts that energy-capital

Energy and the Macroeconomy

substitutability will show up only in the long run in response to a permanent (not transitory) energy price shift. This model should be understood to be a long-run model consisting of three multi-year periods spanning a long lasting, but not permanent, energy price shift far greater than any capital price shift.

As in the putty-clay model, the amount of energy used per unit output may be expressed as a function of the amount of utilized capital, according to the following formula:

$$q_{ei} = b_i q_{ki} \tag{2}$$

where: b_i = the technical coefficient for technology i which relates the amount of energy used per unit output to the amount of utilized capital

Substituting Equation (2) into Equation (1) we obtain

$$R_{rt} = P_k q_{ki} + P_{ert} b_i q_{ki}$$

A. The Initial Equilibrium (Period 1)

Consider the generalizable relationship among technologies with different rates of energy utilization per unit of capital, b_i. Separate firms in a region (say region 1) may select different technologies but (assuming intraregional product transport costs to be zero) must produce output at the same cost per unit, namely R_{11}. Their costs per unit of production may therefore be set equal to one another, as follows:

$$R_{11} = P_k q_{ki} + P_{e11} b_i q_{ki} = P_k q_{kj} + P_{e11} b_j q_{kj}$$

The ratio of capital utilization per unit output by any two separate firms may be determined to be:

$$\frac{q_{kj}}{q_{ki}} = \frac{P_k + P_{e11} b_i}{P_k + P_{e11} b_j} \tag{3}$$

Since the prices of capital and energy in this case are the same, then the firm with the lower rate of energy intensity, b, uses the greater amount of capital per unit [Equation (3)]. At this point in the construction of the model, there is no reason for any region 1 firm to have discriminated in its choice of technology. An illustration of a multi-technology equilibrium in region 1 is shown in Figure 1. The numerical values estab-

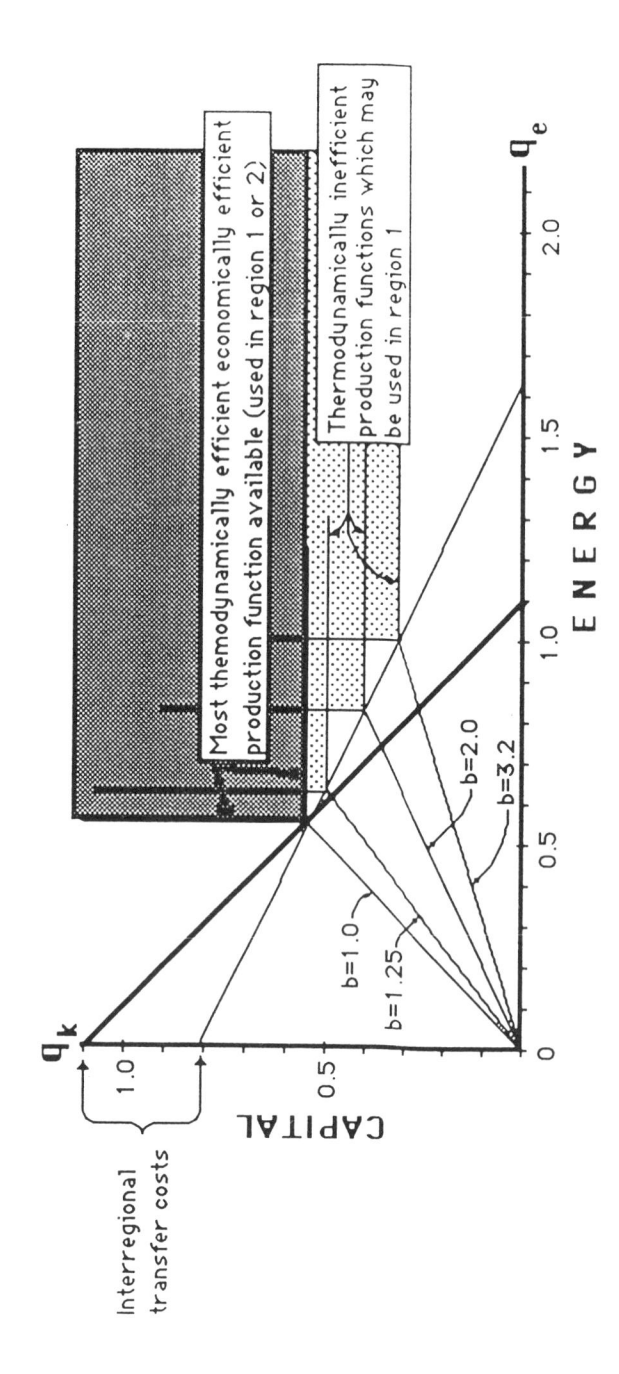

Figure 1. An Illustration of a Period One Equilibrium Situation

Note: See text for specifics of the illustrated case.

Energy and the Macroeconomy 113

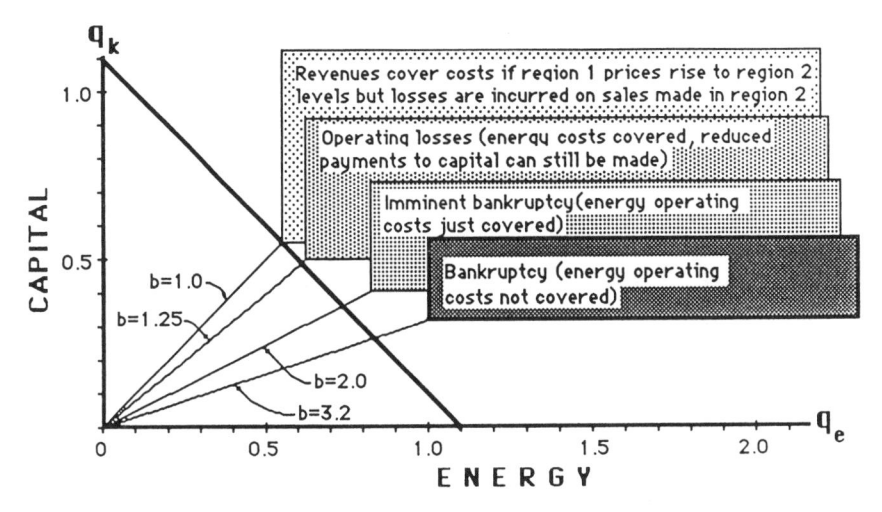

Figure 2. Effects on Region 1 Firms Arising from an Increase of Region 1 Energy Prices to the Level of Region 2 Prices

lishing Figures 1 and 2 values are available in the more lengthy 1986 version of this paper (Santini, 1986b).

If a set of technologies compatible with a given energy price is established in equilibrium, then a move of energy prices away from the equilibrium conditions will require development of new technologies or reevaluation of existing technologies. For a given set of technologies established for a prior energy price, a lowering of energy price will make the least thermodynamically efficient of the technologies (the one with the highest b) the most economically efficient. An increase in price, which is of interest here, causes the most thermodynamically efficient of the technologies to be the most economically efficient (Figure 2). An increase in energy price will also cause the cost of production to be higher, though the increase will be least for the most thermodynamically efficient technology.

In this model a second region with a higher energy price is added to allow the study of capital-energy substitution in cross section in equilibrium. It is assumed that technological knowledge is perfectly mobile, so that any technology that may be used in region 1 may also be freely used in region 2. For any set of technologies that coexist in region 1, only one will be selected as most efficient for use in region 2, so long as region 2 prices differ from region 1 prices. Consequently, it will be possible for interregional indifference among technologies to occur for only one particular technology. When the region 2 energy price is the greater, this must be the most efficient technology. The product costs in region 2 must be higher than in region 1, deterring establishment of production there. However, product transportation costs might be high enough to cause production in region 2 to be desirable. Labeling the most efficient technology as technology a, we may write:

$$P_k q_{ka} + P_{e21} b_a q_{ka} = P_k q_{ka} + P_{e11} b_a q_{ka} + T_{12} \qquad (4)$$

where: T_{12} = product transportation costs between region 1 and region 2.

If we decompose P_{e21} into a part equal to P_{e11} and a remainder equal $P_{e21} - P_{e11}$ and substitute into Equation 4, it can be determined that

$$T_{12} = (P_{e21} - P_{e11}) b_a q_{ka}$$

establishes an equilibrium situation in which all technologies producing in region 1 can sell competitively in both regions, while technology a (only) may produce and sell in region 2. This situation is illustrated in Figure 1.

In an examination of the abstract model results, Santini (1986b) showed that the percent increases in product costs due to transportation from region 1 to 2 are always less than the assumed region 1 to 2 percent increase in energy prices. If it is assumed that region 1 is the energy producing region (it is difficult to imagine otherwise), it can be argued that the model simulates the behavior of a world where the regional variation in prices of energy inputs such as coal, natural gas, electricity, and oil are relatively greatly influenced by transportation costs, when compared to products such as locomotives, ships, automobiles, airplanes, steel, aluminum, copper, and so forth. These are readily defensible results.

Reviewing the technologies allowed when there is regional variation in energy prices, we find that the technology adopted in the region where energy prices are higher is more thermodynamically efficient and uses more capital per unit of output than the average technology in the low energy cost region. The price elasticity of energy for capital is positive, indicating that capital and energy will be estimated to be substitutes in cross sectional regression models which include observations from regions with different energy prices. Further, because of the higher efficiency in the higher price region less energy is consumed per unit of output there than in the low energy price region. Thus, the own price

Energy and the Macroeconomy

elasticity of energy is negative, as it should be, while an increase in energy price results in use of more efficient energy technologies. These results are perfectly consistent with Gibbons' summaries of prior cross section findings but, like his analysis, they do not explain the evidence from time series analysis that the cross price elasticity of capital for energy is negative.

B. The Disequilibrium (Period 2)

The equilibrium analysis dealt with what should be found at a single point in time with a stock of existing technologies, for a given set of regional energy prices. Time series, however, involve adjustments of the stock of existing technologies. This model will be extended to examine what happens when an energy price shock causes an adjustment of technologies from one optimal equilibrium state to another. The analysis focuses on the steps necessary to make the transition. The first term of Equation (1) can be expressed as follows:

$$k_i = P_k q_{ji}$$

where: the rate of spending on capital, k_i, can be said to be an obligation assumed by the firm using technology i.

Suppose that there is an energy price shock to the system which defines the transition from the initial period of equilibrium to the transition period during which the system adjusts toward a new equilibrium. We assume that this shock first occurs in region 1 in such a fashion that it brings P_{e12}, the price of energy, up to the price previously prevailing in region 2. Thus $P_{e12} = P_{e21}$. If the prices are set at the cost of the minimum cost producer (a), the product price will now be higher than before by an amount equal to $(P_{e12} - P_{e11}) b_a q_{ka}$. However, the ability of less efficient competitors to meet capital payment obligations incurred in period 1 will be "squeezed" on the input (variable costs) side by their inherent inefficiency and on the product price side by the ability of their more efficient competitor to sell profitably at a price lower than their cost of production. This energy squeeze is a result of both the energy price increase and the putty-clay nature of capital. It creates an inability of thermodynamically inefficient technologies to repay debt or, alternatively, to internally finance new capital formation.

Capital payment possibilities (K_{pi}) for technology i producing in region 1 and selling in region r are determined by the following formula:

$$K_{pi} = (P_k q_{ka} + P_{e12} b_a q_{ka}) - P_{e12} b_i q_{ki} - T_{rr}$$

The ability of a region 1 firm to repay debt can be estimated by the ratio K_{pi}/k_i, which may be termed a capital coverage ratio. Figure 2 illustrates the possible effect of the energy price increase on region 1 producers. Sharp declines in capital coverage ratios can occur for inefficient producers (see examples in Santini, 1986b). As bankers well know, the principles of operation of the firm require that payments to variable inputs be made first, with the residual, or quasi-rent going to assets fixed in the short run.

This creates a dilemma. While the change in energy price has made the more capital intensive technology a desirable, the fixed (in the physical sense) nature of existing assets has forced more payment to energy and less to capital. Thus, at the very time that capital becomes more desirable, the region 1 firms' ability to purchase it is reduced. Standard microeconomic theory says that the firm should go out of business when $P_{e12} q_{ei} > R$ but stay in business otherwise. If region 1 firms go out of business, the total sales of capital in this system will drop by an amount equal to the prior purchases by those firms. If the more normal situation prevails and the firms stay in business, then the firms will have to greatly curtail their annual purchases of capital and extend the lifetime of capital assets that they currently own.

It has now become desirable to purchase technology a when the next lump of capital is purchased. While the capital coverage ratio illustrates the inability of firms to meet prior obligations, it understates their inability to finance future needs. The most desirable technology now requires more payments to capital than previously assumed by all firms other than owners of technology a ($k_a > k_i$ if $i \neq a$). Consequently, the ability to purchase new capital is even less than the ability to repay prior obligations. If the ability to acquire technology a is defined as A, then:

$$A = K_{pi}/k_a < K_{pi}/k_i \qquad \text{if } i \neq a$$

This exercise shows that, when a firm must adjust from one technology to another, more efficient technology as a result of an energy price shock, the fixed nature of existing capital assets forces a reduction in spending on capital assets in general and a delay in the acquisition date for future capital assets. The greater the energy price increase, the lower the flow of spending on capital assets and the greater the delay in purchasing of replacement assets.

While it is theoretically possible that price increases for the energy intensive goods analyzed here could cause a flow of capital into this sector of the economy, such an argument falls flat on its face when examined empirically. The effect that has been simulated here has been shown in spite of the fact that all firms have been assumed to continue to produce and none of the thoroughly documented declines in product demand for such goods occurs when prices are increased. In view of the sharp declines in sales that occur when prices of energy intensive goods are increased (Lareau and Darmstadter, 1982), it is very doubtful that capital will suddenly flow into this sector under period 2 disequilibrium conditions.

This portion of the disequilibrium period model results in a prediction that a time series analysis including period 1 and period 2 would lead to an *estimate* of a negative cross price elasticity of capital output in response to an energy price increase. The delay and decline in purchase of new capital assets has actually been shown to increase as the need to substitute capital for energy increases. In equilibrium (cross section) such an empirical observation would imply that energy and capital are complements. However, if enough price shock induced disequilibria were present in a time series, the finding of a negative cross price elasticity would nevertheless be perfectly consistent with long run substitutability of capital and energy as shown here. The short run disequilibria, in conjunction with the putty-clay nature of the capital stock, would cause the negative cross price elasticity estimate, but would not imply complementarity.

In this model an explanation of cross section estimates of substitutability of capital for energy consistent with time series estimates of complementarity is developed. Although Berndt and Wood (1979) assert that it is not possible to obtain energy-capital complementarity with a two input model, a two input model has been constructed which shows that the putty-clay nature of capital and the intention of firms to substitute capital for energy after an energy price shock could indeed lead to the econometric illusion that energy and capital are complements. This model does not, however, contradict (nor support) Berndt and Wood's 1979 evidence for complementarity of energy-capital composites for other goods, since other goods are not modeled.

In spite of the fact that this model explains the apparently contradictory cross-section substitutability and time series complementarity findings for energy and capital, the findings remain, in Gibbons' terms, "gloomy." The implication is that any time there is an energy price shock, there will follow some degree of temporary reduction of purchases of energy intensive capital assets. The macroeconomic importance of this problem depends on its pervasiveness in the economy and on other multiplicative effects well known to economists. The empirical evidence cited earlier indicates that the problem is both generic and macro in nature.

IV. DEFINITION OF INNOVATION

For purposes of the depression case discussion, it is necessary to precisely define what is meant by innovation. Economists and engineers can easily talk past one another when discussing innovation because they have different paradigms for the word. By an economist's definition innovation cannot occur without resulting in product cost reduction. By this definition the movement of a product isoquant toward the origin is necessary to have innovation. In Figure 3 the movement from point A to

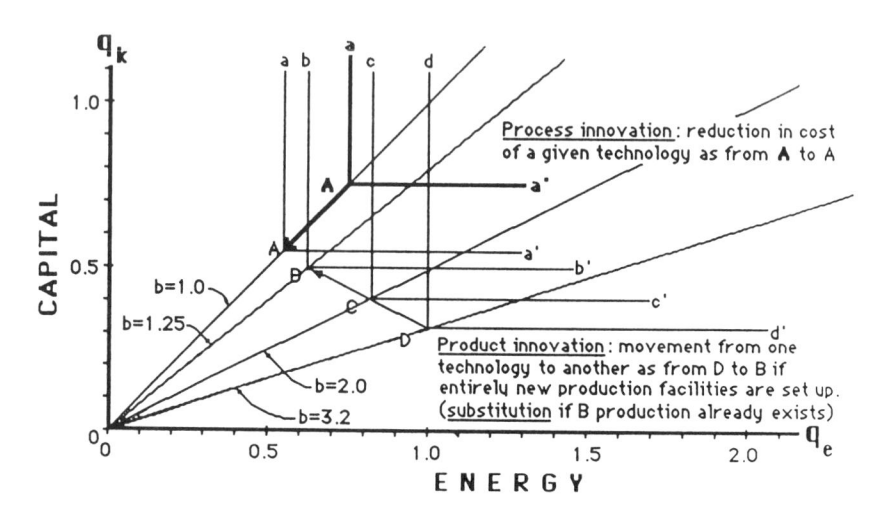

Figure 3. Illustration of Process (Economist's) Innovation and Product (Engineer's) Innovation Definitions

A represents innovation. While the economist may mention substitution of one production technique for another when discussing innovation, when asked to define innovation rigorously the economist is likely to draw a graph of a global production function with capital and labor used as the inputs on the graph's axes and then assert that movement of the production function toward the origin represents innovation. Engineers, however, think of this as process innovation. It is generally accomplished after a new production technique has been adopted and cost reducing modifications of the basic process have been discovered. Process innovations often have the nature of being undiscoverable until large scale production facilities have been set up.

If pressed, the engineer will agree that the reduction of product cost defines successful innovation. However, when writing about or discussing historically significant innovations the engineer will most likely focus on the timing and nature of substitution of one product or production technique for another. As economists know, such substitutions actually take place at the margin. In other words, the actual substitution of one product for another occurs at the point where their product costs have just become equal. Consequently, the engineer's most common conceptualization of innovation is a product-cost-neutral substitution of one production technique (or product) for another. In Figure 3 it is represented by a movement from point D to point B. Among engineers who make the distinction, such shifts are termed product innovation.

It seems intuitive that, where exhaustible resources such as energy are concerned, the movement from a less to a more efficient technology holding current product costs constant represents a cost reduction to society in the long run. Thus, when capital and an exhaustible resource are traded off against one another based on current prices, a resource conserving product cost-neutral substitution away from energy and toward capital should be defined as an innovation in the economic sense. Such an argument could not be made for the trade-off between capital and labor because of the historically documented ability of labor to expand its supply.

V. THE DEPRESSION CASE

A key point concerning the process of introducing a new energy technology is that the introductory sequence will involve product innovation first and accelerated process innovation second. It is reasonable to assume

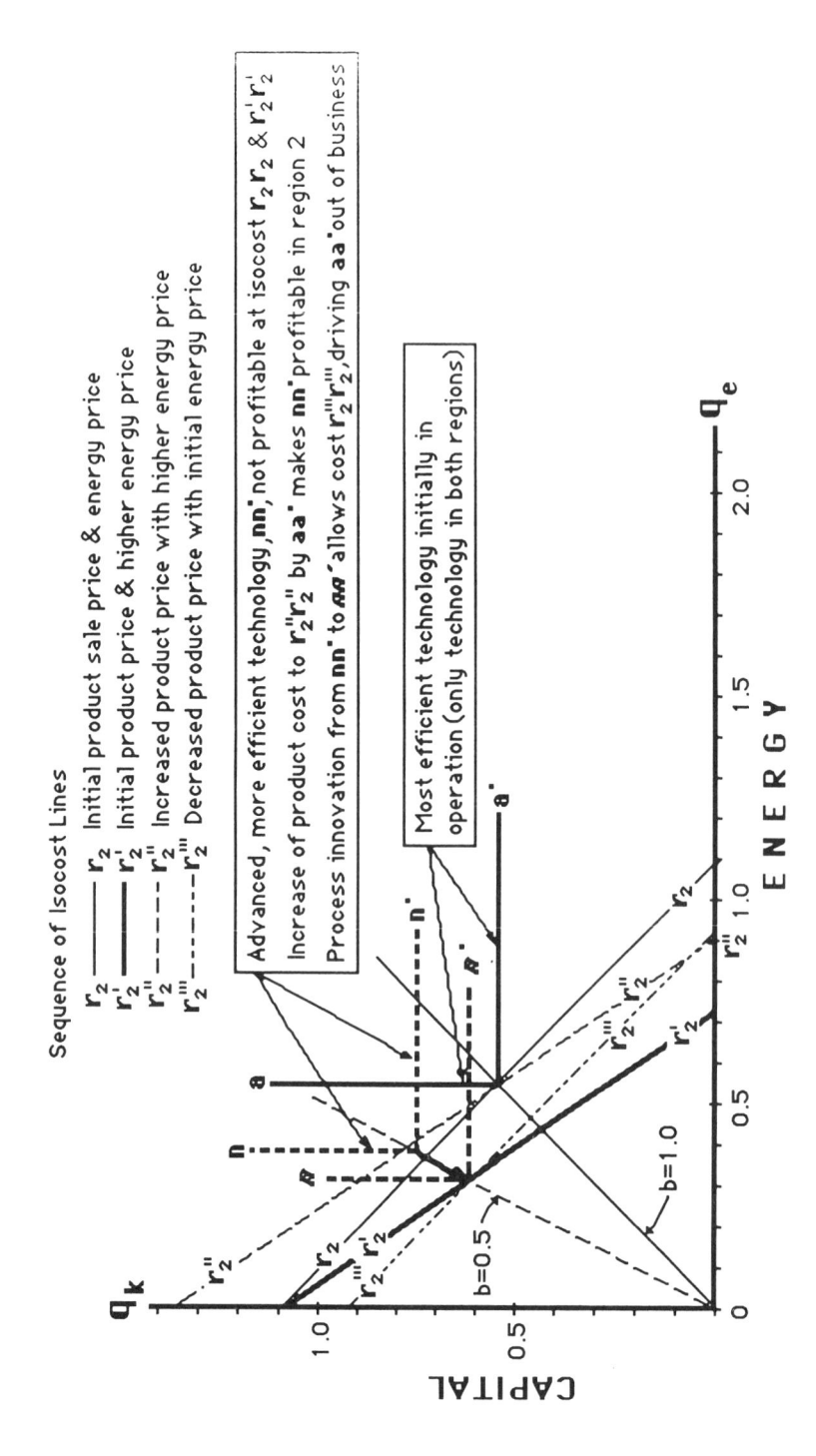

Figure 4. Depression Case: Energy Price Induced Substitution of a New Technology in the High Energy Cost Region

that research and development efforts to reduce the cost of a technically feasible technology would have been used to move a substitute technology near to economic feasibility. Nevertheless, there is likely to be an accelerated process innovation learning curve after product introduction.

The engineering history of transportation technologies is one of long term improvement in thermodynamic efficiency occurring through discrete jumps when significant inventions are introduced. This is equivalent to an intermittent counterclockwise rotation of the b values in Figures 1 and 2. In studying U.S. depressions and periods of relative stagnation, I have observed that, within the transportation sector, the level of thermodynamic efficiency of motor vehicles jumps significantly (Santini, 1985c,d, 1988), while the rate of sales of vehicles drops dramatically. This applies whether the vehicles are sold primarily to individuals (passenger cars) or businesses (locomotives, ships, aircraft). In the context of this model, the average b value for vehicles jumps dramatically during depressions and stagnations, but remains relatively constant in intervening periods. The model developed here implies that the greater the change in b when moving from an old to a new, more thermodynamically efficient technology, the longer and deeper is the decline in capital spending. A characteristic of U.S. depressions is that prior, unusually severe energy price shocks induce the entry of transportation technologies with higher b values. Figure 4 provides an indication of how this process can occur if energy prices in both regions rise. A previously uneconomic capital intensive but highly energy efficient technology becomes attractive as a result of the price rise. In the depression case the entry of the technology is followed by rapid learning and process innovation pushing the costs down. Even though, as has been discussed, energy price collapses often follow the introduction of the new technology, the rapid process innovation for the new technology keeps it the winner, potentially forcing an extension of the period of capital expenditure delay even after energy prices drop.

Graphical and summary empirical details of this process are available in the 1986 paper from which this paper is derived (Santini, 1986b), and further empirical details are available elsewhere (1987b).

Recessions are not characterized by the widespread introduction of new energy efficient transportation technologies, appearing instead to involve relatively short delays in capital spending, with minor adjustments among currently available technologies.

VI. CONCLUSION

The model presented offers one explanation of variation in capital spending during recessions and depressions. The key feature of the model involves delays in spending made necessary by energy cost increases contributing to lower capital coverage ratios and by the need to accumulate more capital before the next planned purchase can take place. Recessions are distinguished from depressions in terms of the degree of improvement of energy efficiency which is desirable, given available energy technologies. The depression case is argued to be one where newly available highly energy efficient (but more capital intensive) technologies extend the period of delay.

The model is set up as a two-region model. By doing so it is implied that one should empirically obtain estimates of substitutability of capital and energy in cross-section tests. In contrast, the time-series model of recessions and depressions is consistent with empirical estimates implying that energy and capital are complements, even though the process described is one of purposeful delay of current capital spending to accomplish an eventual substitution of more capital for energy. These results lend support to Gibbon's conclusions that energy and capital are indeed substitutes in the long run, while implying that short run time series estimates of complementarity are statistical illusions.

ACKNOWLEDGMENTS

This paper is authored by a contractor of the U.S. government under contract No. W-31-109-ENG-38. I would like to thank the Office of Transportation Technologies, the Office of Assistant Secretary for Conservation and Renewable Energy, and the Office of Policy, Planning and Analysis, each within the U.S. Department of Energy, for partial past support of this research. Any opinions expressed here are my own and do not necessarily reflect the views of the U.S. Department of Energy or the Argonne National Laboratory.

REFERENCES

Berndt, E.R., and D.O. Wood. 1975. "Technology, Prices and the Derived Demand for Energy." *Review of Economics and Statistics*, 57: 259–268.

Berndt, E.R., and D.O. Wood. 1979. "Engineering and Econometric Interpretations of Energy-Capital Complementarity." *A.E.R.*, 69: 342–354.

Gibbons, J.C. 1984. "Capital-Energy Substitution in the Long Run." *The Energy Journal,* 5: 109–118.

Gibbons, J.C. 1985. "Energy Prices and Capital Obsolescence: Evidence From the Oil Embargo Period." *The Energy Journal,* 5: 29–43.

Greenhill, B. 1980. *The Ship: The life and death of the Merchant Sailing Ship, 1815–1965.* London: Her Majesty's Stationery Office.

Hamilton, J.D. 1983. "Oil and the Macroeconomy since World War II." *Journal of Political Economy,* 91: 228–248.

Hayek, F.A. (1972) 1979. *A Tiger by the Tail: The Keynesian Legacy of Inflation.* San Francisco: The Cato Institute.

Hotelling H. 1931. "The Economics of Exhaustible Resources." *Journal of Political Economy,* 39: 137–75.

Jevons, W.S. (1865) 1965. *The Coal Question: An Inquiry Concerning the Progress of the Nation, and the Probable Exhaustion of Our Coal-mines.* Edited by A.W. Flux. New York: Augustus M. Kelley.

Lareau, T. J., and J. Darmstadter. 1982. "Energy and Consumer Expenditure Patterns." *Annual Review of Energy,* 7 : 261–292.

Loungani, P. 1986a. *Oil Price Shocks and the Dispersion Hypothesis, 1900–1980.* University of Rochester, Department of Economics, Working Paper No. 33, Department of Economics, University of Rochester, Rochester, New York.

Loungani, P. 1986b. "Oil Price Shocks and the Dispersion Hypothesis." *The Review of Economics and Statistics,* 68, 3: 536–539.

Mork, K.A., and R.E. Hall. 1980. "Energy Prices, Inflation, and Recession, 1974–75." *The Energy Journal,* 1: 31–63.

Rosenburg, N. 1982. *Inside the Black Box: Technology and Economics.* Cambridge: Cambridge University Press.

Santini, D.J. 1983a. "A Discussion of the Indicators Used in Developing an Early and Accurate Judgmental Prediction of Weak Recovery or Depression." Paper presented at the Third International Symposium on Forecasting. Philadelphia, PA, June 5–8.

Santini, D.J. 1983b. "Energy Related Innovation: The Missing Link? A Revision and Integration of the Business Cycle Theories of Keynes and Schumpeter." Paper presented at the 16th International Atlantic Economic Society Conference, Philadelphia, PA, October 6–9.

Santini, D.J. 1983c. "Energy, Innovation and the Business Cycle: Proof and Policy." Pp. 299–312 in *Alternative Energy Sources VI: Proceedings of the 6th Miami International Conference on Alternative Energy Sources.* Washington, DC: Hemisphere Publishing.

Santini, D.J. 1984. "Watch Auto Technology, Fuel Prices to Forecast Slumps." *The Journal of Business Forecasting* (Fall): 18–22, 26.

Santini, D.J. 1985a. "The 'Energy-Squeeze' Model: Energy Price Dynamics in U.S. Business Cycles." *International Journal of Energy Systems,* 5(1): 118–125.

Santini, D.J. 1985b. "Unbalanced Recovery in Energy Related Markets: Growth Retarding Implications for 1985." Paper presented at the Fifth International Symposium on Forecasting, Montreal, Canada, June 9–12.

Santini, D.J. 1985c. "Commercialization of Major Efficiency Enhancing Vehicular Engine Innovations: Past, Present and Future Microeconomic Considerations." *Transportation Research Record,* 1049: 24–34.

Santini, D.J. 1985d. *Direct and Indirect Costs of New Engines.* (Informal Report ANL-EES-TM-156.) Argonne, IL: Argonne National Laboratory.

Santini, D.J. 1986a. "Interactions of Energy and the Macroeconomy—95 Years of Evidence." Pp. 437–442 in *Proceedings of the International Association of Energy Economists North American Conference.* Cambridge, MA.

Santini, D.J. 1986b. "Energy and the Macroeconomy: Capital Spending After an Energy Cost Shock." Paper presented at the 33rd North American Regional Science Association Meetings, Columbus, OH, November 14–16.

Santini, D.J. 1987a. *Verification of Energy's Role as a Determinant of U.S. Economic Activity.* (Formal Report ANL/ES-53.) Argonne, IL: Argonne National Laboratory.

Santini, D.J. 1987b. "Micro and Macroeconomic Responses to Energy Price Shocks: A Discussion of Past, Present and Future Economic Problems in Simultaneously Introducing Alternative Fuels and Major Efficiency-Enhancing Vehicular Engine Innovations." Paper presented at the ORSA/TIMS Joint National Meeting, St. Louis, MO, October 25–28.

Santini, D.J. 1988. "A Model of Economic Fluctuations and Growth Arising From the Reshaping of Transportation Technology." *The Logistics and Transportation Review,* 24, 2: 121–151.

Santini, D.J. 1989a. "The Slowdown of 1989–90: Further Evidence for the Use of Transportation Sector Indicators to Predict GNP Growth." Paper presented at the Fourth Annual Conference on Business Forecasting, September 27–28.

Santini, D.J., and F.J. Belak. 1985. "The Energy Squeeze Model: Further Tests of the Hypothesis." Pp. 32–35 in *Proceedings of the IASTED International Conference, Energy, Power and Environmental Systems.* Anaheim, CA: ACTA Press.

Simon, J.L. 1981. *The Ultimate Resource.* Princeton, NJ: Princeton University Press.

Smith, Adam. (1776) 1976. *The Wealth of Nations.* Chicago: University of Chicago Press.

Tatom, J.A. 1981. "Energy Prices and Short-Run Economic Performance. *Federal Reserve Bank of St. Louis Review,* 63: 3–17.

REDUCING U.S. CARBON DIOXIDE EMISSIONS:
THE COST OF DIFFERENT GOALS

Dale W. Jorgenson and Peter J. Wilcoxen

I. INTRODUCTION

The possibility that carbon dioxide emissions from fossil fuel combustion might lead to global warming through the greenhouse effect has emerged as a leading international environmental concern. Many nations, including the United States, are considering policies to reduce carbon dioxide emissions. Moreover, multilateral action is being discussed under the auspices of the Intergovernmental Panel on Climate Change. For the most part, however, public debate has focused on a few fairly arbitrary targets, such as holding carbon dioxide emissions constant or reducing emissions by 20 percent. Little attention has been devoted to deciding what the optimal target would be.[1] Finding the optimal target requires an accurate assessment of both the costs and benefits of different policies.

In this paper we present a detailed model of the U.S. economy and use it to compute the costs of attaining three different emissions goals by imposing taxes on the carbon content of primary fuels.[2] The goals we consider differ considerably in stringency, so our results give a clear picture of the cost curve lying behind different levels of emissions reductions. We find that costs rise very rapidly, so it is imperative that policy makers carefully assess the benefits of carbon dioxide abatement before adopting a particular target.

The greenhouse effect comes about because several trace gases in the atmosphere, often called greenhouse gases, are transparent to visible light but reflect infrared. Sunlight passes through such gases unimpeded and is absorbed by objects on the ground. Later, much of that energy is re-emitted as infrared radiation. Since greenhouse gases reflect infrared, they tend to trap energy in the atmosphere, which results in heating. The concentration of greenhouse gases in the atmosphere determines how much energy is trapped and thus how much heating occurs.[3]

Carbon dioxide (CO_2) is the most important contributor to the greenhouse effect, although other gases are also important. These other gases include methane (CH_4), chlorofluorocarbons (CFCs), ozone (O_3), Nitrous Oxide (N_2O), and other oxides of nitrogen (NO_x). Much of the carbon dioxide in the atmosphere originates as a natural consequence of respiration, but combustion, particularly of fossil fuels, has increased the atmospheric concentration by 25 percent since the industrial revolution (Schneider, 1989). At present rates of emission, carbon dioxide accounts for about half the increase in the concentration of greenhouse gases, while other gases account for the remainder (Houghton and Woodwell, 1989).

That greenhouse gases trap energy and lead to heating of the atmosphere is not controversial; however, a great deal of uncertainty exists about how much heating will be produced by a given increase in greenhouse gases, and when the heating will occur. Current research indicates that the concentration of greenhouse gases is likely to double sometime in the next century, with mean surface temperatures rising by 1.5 to 5.5 degrees centigrade.[4] Historical data indicate that global temperatures have risen by 0.5 degrees centigrade during the past 100 years, and that the rate of increase has been accelerating.[5]

The environmental consequences of global warming could be severe. It might change patterns of precipitation, cause land to be inundated by increases in the sea level, or increase the frequency of violent storms such as hurricanes. As a consequence, a sizable constituency has developed

Reducing U.S. Carbon Dioxide Emissions

for policy measures to reduce greenhouse gas emissions. For example, the Toronto Conference on the Changing Atmosphere, held in June 1988 and attended by representatives from 48 nations, recommended that world carbon dioxide emissions be reduced to 20% below 1988 levels by the year 2005, and eventually to 50 percent of 1988 levels (National Resources Defense Council, 1989). In the United States, Senator Timothy Wirth has introduced legislation that would sharply reduce carbon dioxide emissions.[6] In addition, environmental groups such as the Natural Resources Defense Council have called for strong, immediate action to reduce the emission of greenhouse gases.[7]

One of the policies proposed for fighting the greenhouse effect is a tax on the carbon content of fossil fuels.[8] This is known as a "carbon" tax, and it could be an effective way to reduce CO_2 emissions. For example, a carbon tax would lead to substitution of other inputs for fossil fuels, and to an increase in the use of fuels such as natural gas that have lower carbon content and hence contribute less to the greenhouse effect. In this paper we examine the costs of a carbon tax in detail, focusing in particular on how those costs change as CO_2 emissions goals become more stringent.

Ours is by no means the first study of greenhouse abatement policies. An important series of studies of the effect of CO_2 restrictions on the U.S. energy sector was initiated by Edmonds and Reilly (1983, 1985).[9] The Edmonds-Reilly approach uses a very detailed model of the energy sector, but it excludes the rest of the economy. Thus, it cannot be used for computing the economy-wide costs of CO_2 abatement, nor can it be used to analyze the impact of restrictions on U.S. economic growth.

Manne and Richels (1990) examined several CO_2 reduction policies using GLOBAL 2100, a five region model of the world economy. GLOBAL 2100 combines a process analysis model of the energy sector with a macroeconomic growth model.[10] The growth model is based on an aggregate production function with inputs of capital, labor, electricity, and nonelectric energy. The production function allows for the possibility that there are "autonomous energy efficiency improvements" which reduce the share of energy in GNP over time. The energy submodel is fairly detailed, including ten electric generation technologies and six sources of nonelectric energy. After examining a number of scenarios combining different assumptions about when various technologies become available, Manne and Richels conclude that a U.S. policy of reducing carbon emissions to the 1985 rate by the year 2000 and subsequently to 80 percent of that rate by 2020 would lower annual GNP by 5 percent by the year 2030. The carbon tax needed to achieve this drop in emissions is enormous, varying over time between $350 and $800 per ton of carbon.[11]

Nordhaus (1989) has recently assessed several policies for controlling greenhouse emissions using rough estimates of the costs and benefits of each.[12] Estimates of benefits are taken from EPA (1988), which quantifies the cost of global warming. A number of alternative abatement strategies are considered—controlling CFCs, reducing CO_2 emissions, reforestation, and imposing a tax on gasoline. Comparing estimates of the marginal cost of reducing greenhouse emissions by the equivalent of one ton of CO_2, Nordhaus argues that the optimal reduction in emissions of greenhouse gases could be achieved with a large reduction in CFCs and a comparatively small reduction in carbon dioxide emissions. Reforestation and gasoline taxes are found to be excessively costly in relation to the amount of carbon removed from the atmosphere.

Whalley and Wigle (1990) have used a global static general equilibrium model to determine the effects of various carbon taxes. Their model divides the world into three regions: high income countries, low income countries, and oil exporters. They conducted several experiments, each designed to achieve a 50 percent decrease in the production of energy from sources that contribute to the greenhouse effect. (Underlying this is an implicit assumption that greenhouse gases and certain forms of energy are produced in fixed proportions.) Their results show that world welfare would fall by more than $250 billion annually (U.S. dollars). Depending on how the carbon tax was implemented, the revenue raised could be substantial—from $100 billion to $300 billion dollars a year in the high income countries alone. An interesting feature of the study is the possible distributions of these revenues among the three regions.

From these studies a great deal of valuable information has been accumulated about the economic impact of policies to limit the emissions of greenhouse gases. However, the analysis of the impact on U.S. economic growth of restrictions on these emissions is seriously incomplete. In order to measure this cost it is essential to model the responses of businesses and households at a highly disaggregated level. Policies such as carbon tax are intended to reduce fossil fuel use by inducing producers and households to substitute toward other inputs, and a detailed model is needed to capture these effects. Moreover, carbon taxes are likely to affect the price of new capital goods, and thus will affect the rate of capital accumulation. To capture this effect requires a model with endogenous capital formation. In addition, carbon taxes will increase the price of energy to purchasers, which may reduce or accelerate technical change. To address these concerns requires a disaggregated, dynamic general equilibrium of the U.S. economy in which technical change is endogenous. In the remainder of this paper we present such a model and use it to examine the effects of several carbon taxes.

II. AN OVERVIEW OF THE MODEL

The results presented in Section III are based on simulations we conducted using a disaggregated, econometrically estimated intertemporal general equilibrium model of the United States. The model itself is an extension of our earlier work on environmental regulation, and is documented in detail in Jorgenson and Wilcoxen (1990a, 1990b). Rather than presenting the entire model again, in this section we confine ourselves to outlining a few of its key features and discussing how we extended it to calculate carbon emissions.

A. Producer Behavior

Several of the model's most important features are closely connected to our submodel of producer behavior. For example, production is moderately disaggregated: total output is divided into 35 separate commodities, each of which is produced by one or more of 35 industries. The industries correspond roughly to two-digit SIC classifications, and are shown in Table 1. This level of detail allows us to measure the effect of shocks on fairly narrow segments of the economy. Since carbon dioxide emissions are concentrated in energy production—a small part of the overall economy—a disaggregated model is essential for examining the sectoral effects of global warming policies.

Each of the 35 industries is represented by an econometrically estimated nested translog unit cost function. At the function's top level, output is produced using capital, labor, energy and materials (KLEM). Capital and labor are both primary factors purchased directly from households. Energy and materials, on the other hand, are translog aggregates of intermediate goods. The energy aggregate is composed of inputs of coal, crude petroleum, refined petroleum, electricity and natural

Table 1. Definitions of the Industries

Number	Description
1	Agriculture, forestry, and fisheries
2	Metal mining
3	Coal mining
4	Crude petroleum and natural gas extraction
5	Nonmetallic mineral mining
6	Construction
7	Food and kindred products
8	Tobacco manufactures
9	Textile mill products
10	Apparel and other textile products
11	Lumber and wood products
12	Furniture and fixtures
13	Paper and allied products
14	Printing and publishing
15	Chemicals and allied products
16	Petroleum refining
17	Rubber and plastic products
18	Leather and leather products
19	Stone, clay, and glass products
20	Primary metals
21	Fabricated metal products
22	Machinery, except electrical
23	Electrical machinery
24	Motor vehicles
25	Other transportation equipment
26	Instruments
27	Miscellaneous manufacturing
28	Transportation and warehousing
29	Communication
30	Electric utilities
31	Gas utilities
32	Trade
33	Finance, insurance, and real estate
34	Other services
35	Government enterprises

gas[13] while the materials aggregate is composed of inputs of all other intermediate goods. Minimizing costs subject to this specification allows us to derive factor demands for capital, labor and intermediate inputs of the 35 commodities. When fully parameterized, these demands completely describe producer behavior.

To parameterize the producer submodel (and, indeed, the rest of the model) we constructed a special data set of consistent input-output tables running from 1947 through 1985.[14] This data set allowed us to estimate all parameters in all of the model's behavioral equations, a feature which most clearly distinguishes our model from others. This method of parameterization, known as the econometric approach, stands in marked contrast to the calibration method used for most general equilibrium models. Calibration involves choosing the model's parameters so that the model will replicate a particular year.[15] Because it requires fairly little data, calibration has been widely applied in general equilibrium modeling.

By taking the econometric approach, however, we gained several advantages over calibration. First, by using a long time series of data (rather than a single point) we are able to estimate more flexible functional forms. Thus, our approach imposes less structure on the data than the simple functional forms used for calibration. We do not, for example, need to assume that production is Cobb–Douglas or CES. A second advantage is that estimated parameters based on a long time series are less likely to be corrupted by noise. Calibrated parameters, on the other hand, are forced by construction to absorb all noise present in the data. This poses a severe problem when the benchmark year is unusual in some respect because calibrated parameters will build that distortion into the model. Estimation avoids this problem by reducing the influence of any particular year's data on the parameters. Most importantly, however, by estimating each industry's cost function on a consistent set of time series data, our model implicitly incorporates elasticities of substitution consistent with historical observations.[16]

The third important feature of the producer submodel is our treatment of technical change. As part of each industry's cost function, we include several terms to allow for technical change.[17] Most other models used to study global warming, such as Manne and Richels (1990), assume a constant exogenous rate of technical change. In our model, however, technical change is endogenous. Moreover, it is determined at the industry level, which allows different sectors to grow at different rates. In addition, each industry's technical change may be biased toward some inputs and away from others. Differing rates of biased technical change are a common feature of historical data, but are necessarily absent from aggregated models. By including endogenous industry-level technical change, our model is able to capture the medium run evolution of individual sectors much more accurately.

In sum, the salient features of the production submodel are that it is disaggregated and econometrically estimated, and that it allows for industry-level biased technical change. In addition, it fully captures substitution among intermediate inputs. We now turn briefly to a discussion of the model's final demands: consumption, investment, government spending, and foreign trade.

B. Consumption

Our final demand vector giving household consumption by commodity is the end result of a three-stage intertemporal optimization problem. At the first stage, each household allocates full wealth (the sum of financial wealth, discounted future labor earnings, and an imputed value of leisure time) across different time periods according to its rate of time preference and its intertemporal elasticity of substitution. We formalize this decision using a representative agent who maximizes an intertemporal utility function subject to an intertemporal budget constraint. The agent's optimal allocation must satisfy a set of necessary conditions which can be summarized in the form of an Euler equation.[18] The Euler equation is forward-looking, so current consumption, and hence the rate of saving, will depend on expectations about future prices and interest rates. Because capital formation is an important contributor to economic growth, this formulation of the savings decision plays a significant role in the model. We will return to this point in the section on investment.

Once households have allocated full wealth across periods, they begin the second stage of their optimization: deciding on the mix of leisure and goods to consume in each period. As in the intertemporal allocation, we simplify the representation of household preferences between goods and leisure by the use of a representative consumer. The representative consumer has a translog intraperiod indirect utility function which depends on the prices of leisure and an aggregate consumption good. (We take the price of leisure to be the after-tax wage rate and the price of the aggregate consumption good to be a price index based on the commodities consumed.) From this we derive the consumer's demands for leisure and goods in each period as a function of prices and the amount of full wealth allocated to the period. This produces an allocation of the household's time endowment, which is given exogenously, between leisure time and the labor market. Thus, the second stage of the consumer model determines labor supply.

The third stage of the household optimization problem is the allocation of consumption expenditures among capital, labor, and the 35 commodities. At this stage, we abandon the representative consumer assumption and instead follow the methodology of Jorgenson, Lau, and Stoker (1982) by formulating a system of individual household demand systems which can be aggregated. We then distinguish between 672 household types based attributes such as the number of household members and the geographic region in which the household is located. For each of these households, we follow the approach of Jorgenson and Slesnick (1987) by using a nested translog tier structure to represent demands for individual commodities.[19]

As with production, all behavioral equations in all stages of the consumer model are econometrically estimated. This includes the Euler equation, the allocation equations for leisure and personal consumption, and the equations governing the allocation of consumption among commodities.[20] Thus, our household model incorporates historical substitution shown by consumers. Moreover, an important feature of our specification is that we do not require household demands to be homothetic. Thus, as incomes rise the pattern of consumption will shift, even in the absence of price changes. This captures an important and often noted feature of historical data which is usually ignored in general equilibrium modeling.

C. Investment and Capital Accumulation

As already noted, an important feature of the model is that it is based on intertemporal optimization by households, which is the source of our savings supply function. In addition, we also assume intertemporal behavior on the part of investors. Both types of intertemporal behavior are very important for greenhouse abatement simulations as much of the impact of regulation occurs far in the future. Since many of these effects will be anticipated by households and firms, future events will have consequences for current decisions. Saving, for example, depends on households' expectations of future earnings and interest rates, while investment depends on firms' expectations of future wages and prices. Changes in saving and investment affect the rate of capital accumulation, and hence the rate of economic growth.

Our investment model is based on perfect foresight or national expectations. In particular, we require that the price of new investment goods always be equal to the present discounted value of the returns expected on an extra unit of capital.[21] For tractability, we assume there is a single capital stock in the economy which is perfectly malleable and can be reallocated between industries at zero cost.[22] The total supply of capital, however, is fixed at any time by past investment behavior. This implies that the return on a unit of capital in a given period is precisely equal to the economy-wide rental price of capital goods. In addition, we assume that new capital goods are produced out of individual commodities according to a production function estimated from historical data, so the price of new capital will depend on commodity prices. Thus, the price of capital goods and the discounted value of future rental prices must be brought into equilibrium by adjustments in the term structure of interest rates. Finally, the quantity of investment done in each period is determined by the amount of savings made available by households.

The production function for new capital goods was estimated using final demand data for investment over the period 1947–1985. Thus, our model incorporates substitution between different inputs in the composition of the aggregate capital good. This feature sometimes plays an important role. In our earlier work on environmental regulation, for example, we found that a substantial drop in the price of automobiles would shift investment toward motor vehicles and away from other durable goods.

In sum, capital accumulation is the outcome of intertemporal behavior on the part of households and firms. Households determine the amount of savings available in each period through intertemporal utility maximization. Firms, for their part, invest until the returns on additional investment are driven to the cost of new capital goods. Finally, savings and investment are equilibrated by the interest rate.

D. Government and Foreign Trade

The two remaining final demand categories are the government and the foreign sector. Beginning with the government, we determine final demands for government consumption from the income–expenditure identity for the government sector. The first step is to compute total tax revenue by applying exogenous tax rates to appropriate transactions in the business and household sectors. We then add the capital income of government enterprises (determined endogenously) and nontax receipts (exogenous) to tax revenue to obtain total government revenue.

Next, we make a important assumption about the government budget deficit; namely that it can be specified exogenously.[23] We add the deficit

Reducing U.S. Carbon Dioxide Emissions

to total revenue to obtain total government spending. To arrive at government purchases of goods and services, we subtract interest paid to domestic and foreign holders of government bonds together with government transfer payments to domestic and foreign recipients. We allocate the remainder among commodity groups according to fixed shares constructed from historical data. Finally, we determine the quantity of each commodity by dividing the value of government spending on the good by its price.

Foreign trade, on the other hand, has two components: imports and exports. Imports are handled by assuming that they are imperfect substitutes for similar domestic commodities.[24] To implement this, we assume the goods actually purchased by households and firms are translog aggregates of domestic and imported products, where parameters of the aggregation function are determined by estimation. Thus, each commodity is governed by a separate elasticity of substitution between foreign and domestic goods. The result is that intermediate and final demands implicitly determine imports of each commodity. The prices of imports are given exogenously in each period.

Exports, on the other hand, are modeled by a set of explicit foreign demand equations, one for each commodity, which depend on foreign income (given exogenously) and the foreign price of U.S. exports. Foreign prices are computed from domestic prices by adjusting for subsidies and the exchange rate. The demand elasticities appearing in these equations were estimated from historical data.[25]

The final important part of the foreign trade submodel is our treatment of the current account and the exchange rate. Without an elaborate model of international trade (far beyond the scope of this study), it is impossible to determine both the current account and the exchange rate endogenously. Thus, in the simulations reported below, we take the current account balance to be exogenous and the exchange rate to be endogenous.

E. Computing Carbon Emissions

The most important remaining feature of the model is the way in which carbon dioxide emissions are calculated. For tractability, we assume CO_2 is emitted in fixed proportion to fossil fuel use. This implicitly assumes that nothing can be done to reduce the CO_2 produced by any given combustion process, but in practice that is largely the case.[26]

Table 2. Carbon Emissions Data for 1987

Item	Coal	Oil	Gas
Unit of measure	ton	bbl	kcf
Heat content			
(10^6 BTU per unit)	21.94	5.80	1.03
Emissions rate			
(kg per 10^6 BTU)	26.9	21.4	14.5
(kg per unit)	590.2	124.1	14.9
Total domestic output			
(10^9 units)	0.9169	0.3033	17.8
Total carbon emissions			
(10^6 tons)	595.3	414.1	268.6

For comparability with other studies, we measure CO_2 emissions in tons of contained carbon.[27]

We calculated the carbon content of each fossil fuel in the following way. From the Department of Energy we obtained the average heat content of each fuel in millions of BTU per quantity unit (Department of Energy, 1990). Next, we obtained data from the Environmental Protection Agency on the amount of carbon emitted per million BTU produced from each fuel.[28] Multiplying EPA's figures by the heating value of the different fuels gives the carbon content of a unit of each fuel. Total carbon emissions can then be calculated using figures on total fuel production. Table 2 shows data for each fuel in 1987.

Our simulation model, however, is normalized so that all prices are equal to one in 1982. Thus, its quantities do not correspond directly to physical units. Moreover, the model has a single aggregate sector for oil and gas. To convert the figures above into a form appropriate for the model's quantity units, we summed carbon production for oil and gas and divided by the model's output for industry 4 (oil and gas extraction) in 1987. This gave the carbon coefficient for that industry. Similarly, the coefficient for coal was computed by dividing total carbon production from coal by the model's 1987 value for coal output. These coefficients were then used to compute carbon emissions in each simulation. We now turn to a brief discussion of the model's base case.

F. The Base Case

In order to solve the model, we must provide values for all exogenous variables in all periods. We accomplish this in two steps. First, we develop a set of default assumptions about the values each exogenous variable will have over time in the absence of changes in government policy. This is used to generate a simulation called the "base case." The second step is to change certain exogenous variables to reflect a proposed policy and then solve the model again to produce a "revised case." We can then compare the two simulations to assess the effects of the policy. Thus, the assumptions underlying the base case are of some importance in understanding the model's results.

Because the model includes agents with perfect foresight, we must solve it far into the future. In order to do that, we project values for all exogenous variables over the period 1990–2050. After 2050 we assume the variables will remain constant at their 2050 values, which allows the model to converge to a steady state by the year 2100. Some of the most important or interesting projections are noted briefly below; a more detailed discussion appears in Jorgenson and Wilcoxen (1990b).

First, we set all tax rates to their values in 1985, the last year in our sample period. Next, we assume that foreign prices of imports (in foreign currency and before tariffs) remain constant in real terms at 1985 levels. Third, we project a gradual decline in the government deficit through the year 2025, after which the deficit is held at four percent of the nominal value of the government debt. This has the effect of maintaining a constant ratio of the value of the government debt to the value of the national product when the inflation rate is 4 percent (as it is in our steady state). Fourth, we project that the current account deficit will fall gradually to zero by the year 2000. After that we project a small current account surplus sufficient to produce a stock of net claims on foreigners by the year 2050 equal to the same proportion of national wealth as in 1982.

Finally, the most important exogenous variables are those associated with U.S. population growth and the corresponding change in the economy's time endowment. We project population by age, sex, and educational attainment through the year 2050 using demographic assumptions consistent with Social Security Administration forecasts.[29] After 2050 we hold population constant, which is roughly consistent with Social Security projections. In addition, we project educational attainment by assuming that future demographic cohorts will have the same level of attainment as the cohort reaching age 35 in the year 1985. We then transform our population projection into a projection of the time endowment used in our model of the labor market by assuming that relative wages across occupations are constant at 1985 levels. Since capital accumulation is endogenous, these population projections effectively determine the size of the economy in the more distant future.

III. THE IMPACT OF DIFFERENT EMISSIONS TARGETS

We now turn to our results on the effects of using a carbon tax to achieve different CO_2 emissions goals. All together, we ran three simulations in addition to the base case, one for each of the following policies:

1. Stabilizing carbon emissions at the 1990 base level beginning immediately.
2. Decreasing carbon emissions gradually over 1990–2005 until they are 20 percent below the 1990 base level.
3. Doing nothing until 2000, then gradually increasing the carbon tax over 2000–2010 to stabilize emissions at the year 2000 base level.

These policies vary considerably in stringency. In 1990, base-case fossil fuel use produced 1576 million tons of carbon. Policy 1 would keep that level constant forever, even in the face of rapid GNP growth. Policy 2, however, is even more restrictive: it requires emissions to drop to 1,261 million tons by 2005 and remain at that level forever. Policy 3, on the other hand, is the least restrictive: it allows emissions to rise to the base-case year 2000 level of 1,675 million tons.

In each simulation, we constrained total carbon emissions and allowed the level of the carbon tax to be determined endogenously. The tax was applied to primary fuels (industries 3 and 4) in proportion to carbon content. Because even the least stringent policy produces substantial tax revenue, it was also necessary to make an assumption about how the revenue would be used. In these simulations, we held the real value of government spending constant at its base-case level and allowed the average tax on labor to adjust to keep the difference between government spending and government revenue equal to the exogenous budget deficit. At the same time, we held the *marginal* tax on labor constant, so adjustments in the average rate reflect changes in the implicit zero-tax threshold.

A. Long-Run Effects

The principal direct consequence of all three carbon control strategies is to increase purchasers' prices of coal and crude oil. This can be seen most clearly by examining the model's results for each simulation at a particular point in time, so in this section we present detailed results for the year 2020. Our model is most suitable for medium-run analysis (periods of 20–30 years), so for our purposes 2020 is the long run.

We begin with results for the first experiment: holding emissions at 1990 levels. By the year 2020, maintaining 1990 emissions will require a tax of $16.96 per ton of carbon contained in primary fuels.[30] Using the data in Table 2, it can be shown that this amounts to a tax of about $11.01 per ton of coal, $2.32 per barrel of oil, or $0.28 per thousand cubic feet of gas. The tax would generate revenue of $26.7 billion annually.

The rising price of fossil fuels provokes substitution toward other energy sources and away from energy in general. Total BTU consumption falls by 12 percent to about 68 quads. This substitution away from energy, and hence toward more expensive production techniques, results in a drop of 0.7 percent in the capital stock and 0.5 percent in real GNP. These figures are fairly small because they measure, in a loose sense, the welfare losses from introducing a small distortionary tax. Because revenue from the tax is returned to households through essentially lump-sum adjustments in the income tax, social welfare falls purely due to the inefficiency of the tax.

At the commodity level the impact of the tax varies considerably. Figure 1 shows changes in the supply price of the 35 commodities measured as percentage changes relative to the base case. The largest change occurs in the price of coal (commodity 3), which rises by 40 percent. This, in turn, increases the price of electricity (commodity 30) by about 5 percent. Electricity prices rise considerably less than coal prices because coal accounts for only about 13 percent of total utility costs. Other prices showing significant effects are those for crude and refined petroleum (goods 4 and 16) and gas utilities (good 31). These rise, directly or indirectly, because of the tax on oil.

These changes in prices affect demands for the commodities, which in turn determine how industry outputs are affected. Figure 2 shows percentage changes in quantities produced by the 35 industries. Most of the sectors show only small changes in output. Coal mining (sector 3) is the exception: its output falls by 26 percent. Coal is affected strongly

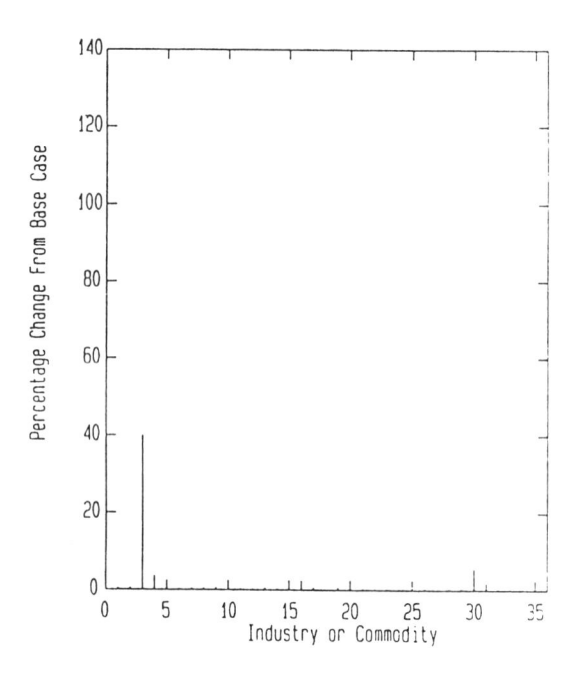

Figure 1. Carbon Tax (1990) — Effect on Prices

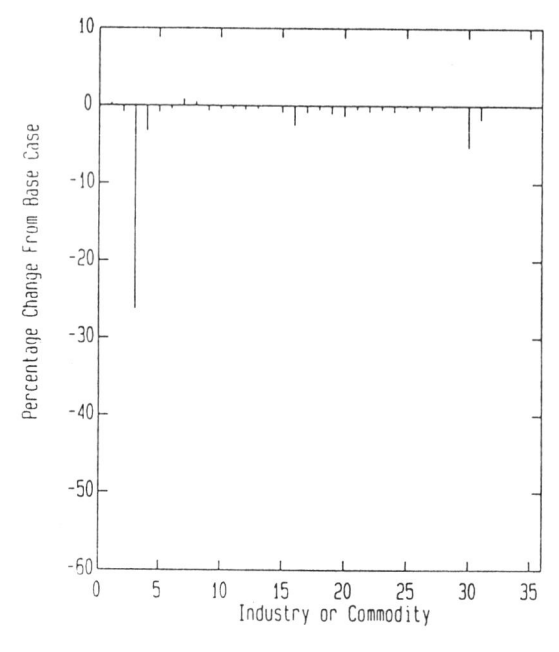

Figure 2. Carbon Tax (1990) — Effect on Output

Reducing U.S. Carbon Dioxide Emissions

because the demand for it is somewhat elastic. Most coal is purchased by electric utilities, which in our model can substitute toward other fuels when the price of coal rises. Moreover, the utilities also have some ability to substitute other inputs, such as labor and capital, for energy, further reducing the demand for coal. Because electric utilities play such an important role in determining how a carbon tax affects coal mining, we now digress briefly to discuss how the utilities are represented in the model.

Electric utilities, like all other sectors, are represented by a nested translog unit cost function. The top tier of the function gives cost in terms of the prices of four inputs: capital, labor, an energy aggregate, and a materials aggregate. Substitution between energy and other inputs takes place at this level. The price of the energy aggregate itself is formed at a lower tier by translog aggregation of the prices of five inputs: coal, crude petroleum, refined petroleum, electricity, and natural gas from gas utilities. Substitution between fuels takes place at that level.

Estimated parameters govern the ease of substitution at both the KLEM and energy tiers of the cost function. At the KLEM level, substitution between energy and capital is very inelastic (an elasticity of –0.15), substitution between energy and labor is moderately inelastic (–0.64), and substitution between energy and materials is slightly elastic (–1.16). Thus, increases in the relative price of energy will, for the most part, induce substitution toward materials. In addition, substitution possibilities also exist at the energy tier. The elasticity of substitution between coal and refined petroleum is –0.7, although between coal and natural gas it is only –0.1. Thus, an increase in the relative price of coal will produce some substitution toward other fuels. Overall, the parameters appearing in the cost function for electric utilities imply that an increase in the relative price of coal will lead to substitution toward other fuels and toward non-energy inputs.

The second policy we considered was a 20 percent reduction below 1990 emission rates, to be phased in gradually over 15 years. By 2020, this would amount to a drop of 32 percent below base case emissions, and would require a tax of $60.09 per ton of carbon. Using the data in Table 2, this is equivalent to a tax of $39.01 per ton of coal, $8.20 per barrel of oil, or $0.98 per thousand cubic feet of gas. The tax would produce $75.8 billion in revenues. Comparing these results to those for maintaining 1990 emissions shows that the tax would more than triple, from $17 to $60. At equilibrium, the tax gives the marginal cost of reducing emissions by an additional ton of carbon, so it is clear that further reductions are becoming significantly more difficult.

Tighter carbon regulations also lead to a reduction in total fossil fuel BTU production to 57 quads, a drop of 27 percent from the base case. This, in turn, reduces the capital stock by 2.2 percent and real GNP by 1.6 percent. These figures are about triple the values obtained for holding emissions at 1990 levels. Although the changes in capital and GNP appear small, recall that they are measures of deadweight loss associated with fairly large marginal changes in the energy sector.

At the commodity and industry level, results for this experiment are qualitatively similar to those for maintaining 1990 emissions, although they are numerically somewhat different. Figure 3 shows percentage changes in commodity prices relative to the base case. The price of coal more than doubles, rising by 137 percent from its base case value. The price of oil rises by 13 percent, while that of electricity rises by about 18 percent. The prices of refined petroleum and natural gas also rise, but by somewhat less. Comparing Figures 3 and 1 shows how this simulation compares with the previous one. In particular, commodity prices rise roughly in proportion to the increase in the carbon tax: the tax rises by a factor of 3.5, and so do most of the percentage changes in commodity prices.

The quantity results, shown in Figure 4, display a similar pattern except that they scale up in proportion to the change in carbon reductions rather than the change in taxes. That is, reducing emissions to 20 percent below 1990 levels requires a cut of about twice the size needed to reach 1990 levels. Thus, percentage changes in quantities from the base case are about twice those of the previous experiment. The most important results are the 53 percent drop in coal production and the 15 percent drop in electricity produced.

In contrast, the looser restrictions implied by maintaining emissions at year 2000 levels produce much smaller effects on the economy. The tax required is only $8.55 per ton of carbon, which implies charges of $5.55 per ton of coal, $1.17 per barrel of oil, or $0.14 per thousand cubic feet of gas. The tax would produce $14.4 billion annually in revenue. Aggregate effects are also considerably smaller than in the two previous scenarios. The capital stock will fall by 0.4 percent, and GNP will drop by 0.3 percent, about half the value obtained in the 1990 simulation. This is quite reasonable since the cut in emissions is about half as deep. The industry results look qualitatively so similar to those of the previous

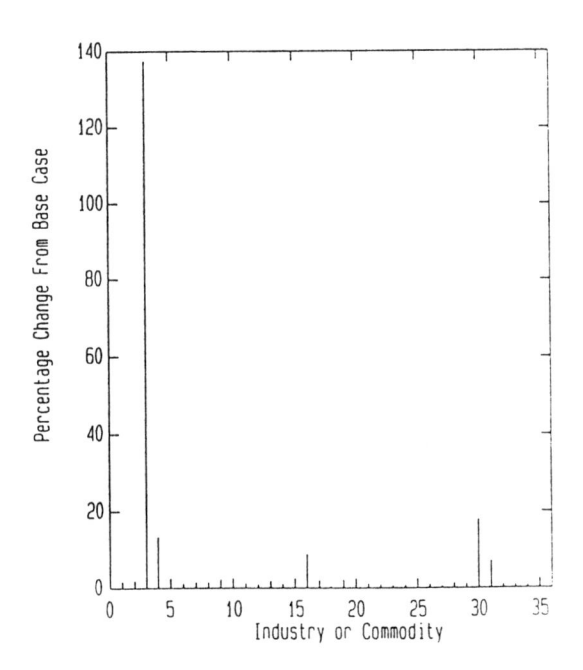

Figure 3. Carbon Tax (80%, 1990) — Effect on Prices

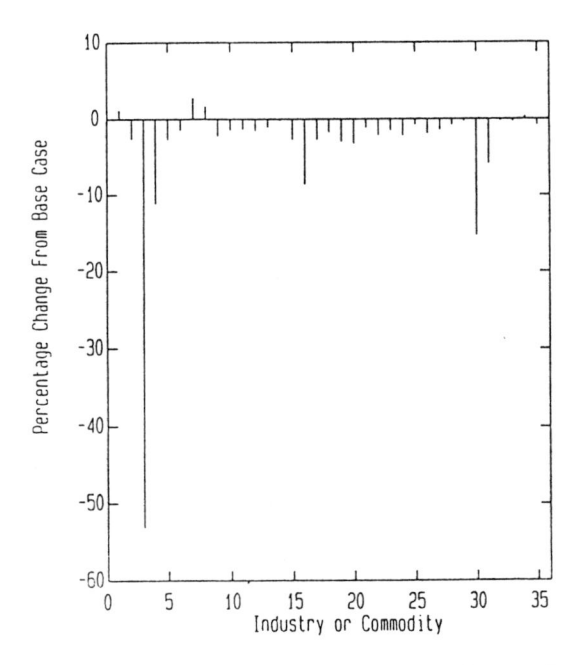

Figure 4. Carbon Tax (80%, 1990) — Effect on Output

Table 3. Summary of Long-Run Carbon Tax Simulations

Variable	Unit	2000 Level	1990 Level	80% of 1990
Carbon emissions	%Δ	−8.4	−14.4	−31.6
Carbon tax	$/ton	8.55	16.96	60.09
Tax on coal	$1/ton	5.55	11.01	39.01
Tax on oil	$/bbl	1.17	2.32	8.20
Tax on gas	$/kcf	0.14	0.28	0.98
Labor tax rate	Δ	−0.25	−0.45	−1.22
Tax revenue	Bil.$	14.4	26.7	75.8
BTU production	%Δ	−7.1	−12.2	−27.4
Capital Stock	%Δ	−0.4	−0.7	−2.2
Real GNP	%Δ	−0.3	−0.5	−1.6
Price of coal	%Δ	20.3	40.0	137.4
Quantity of coal	%Δ	−15.6	−26.3	−53.2
Price of electricity	%Δ	2.9	5.6	17.9
Quantity of electricity	%Δ	−2.9	−5.3	−15.3
Price of oil	%Δ	1.8	3.6	13.3

experiments that we omit the graphs. The principal numerical result is that coal prices rise by 20 percent while coal output shrinks by about 16 percent.

The results of all three carbon tax simulations are summarized in Table 3, in which the policies are listed in order of increasing stringency. From these results it appears that maintaining emissions at the year 2000 base-case level can be accomplished with a very low carbon tax and minimal disturbance of the economy. The strongest effect would be felt by the coal mining industry, which would see its demand fall as electric utilities substituted toward other fuels. More stringent regulations, however, would lead to markedly higher energy prices and greater disruption of the economy. Under any scenario, however, coal mining would bear the brunt of the changes brought about by the tax. Of the remaining sectors, electric utilities would be affected most strongly.

B. Intertemporal Results

Carbon restrictions adopted today will have effects far into the future. At the same time, anticipated future restrictions will have effects today. To assess the intertemporal consequences of carbon taxes, we now turn to the model's dynamic results. As with the long run results, we begin by discussing a carbon tax designed to maintain emissions at 1990 levels. Following that, we examine the dynamic behavior of other experiments.

Reducing U.S. Carbon Dioxide Emissions

The path of the carbon tax needed to maintain 1990 emissions is shown in Figure 5. Base case emissions increase over time, so the tax grows gradually, about $0.70 per year, over the next few decades. It reaches a peak around the year 2020 when our forecast of the U.S. population crests.[31] The tax produces significant reductions in carbon emissions which are shown in Figure 6 as percentage changes from the base case. Emissions begin dropping immediately and by 2020 are about 14 percent below their unconstrained level.

As suggested by the long-run results, the principal effect of the tax is to reduce coal mining. This is shown clearly in Figure 7, which gives percentage changes in coal output from the base case. Production gradually slows as the tax is introduced. It does not, however, fall all the way back to its 1990 level—some of the reduction in emissions comes about through reductions in oil consumption. This can be seen in Figure 8, which gives percentages changes in crude petroleum and natural gas extraction over time.

The increasing price of energy raises costs and reduces household income. This, in turn, changes the rate of capital accumulation. The outcome is shown in Figure 9, which gives percentage changes in the capital stock from the base case. Unlike variables in the preceding graphs, the capital stock does not start declining immediately; instead, it tends to remain near its base-case level for the first few years. This comes about because of intertemporal optimization by households. From a household's point of view, the effect of the tax is to decrease its real income by an amount related to the tax's deadweight loss.[32] Thus, the household regards carbon taxes as reductions in future earnings, so it reacts by lowering consumption in all periods. In the early years, however, the carbon tax is minimal and household income is largely unaffected. During that period, therefore, the drop in consumption leads to an increase in saving. This helps maintain investment—and thus the capital stock—in the early years of the simulation. Eventually, the income effect of the tax begins to be felt and the capital stock finally starts to decline relative to the base case.

The decline in growth of the capital stock leads to a drop in GNP growth, as shown in Figure 10. Over time GNP gradually falls by about half a percent relative to the base case. The capital stock, however, is not the only factor contributing to the decline. In addition, higher energy prices reduce the rate of technical change in industries which are energy-using. This leads to slower income growth and helps keep GNP below

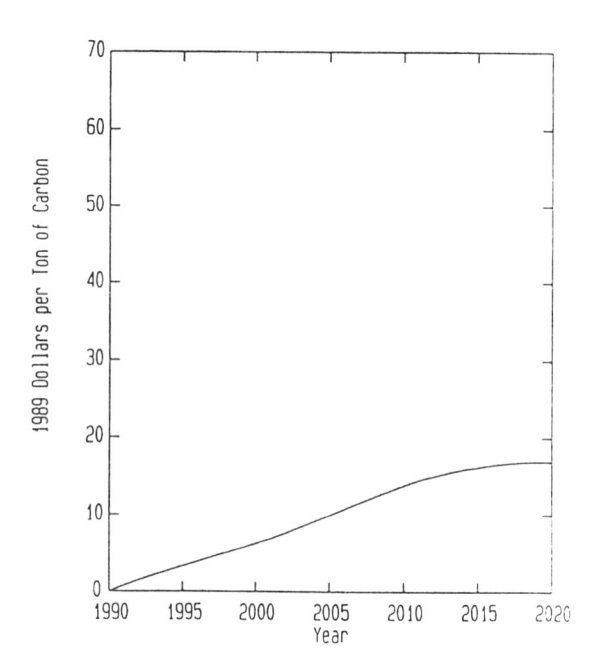

Figure 5. Carbon Tax to Maintain 1990 Emissions

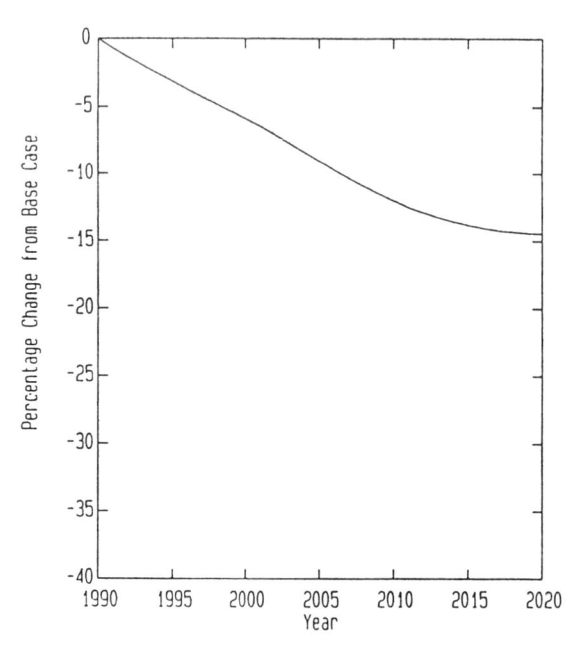

Figure 6. Carbon Emissions Under a Carbon Tax

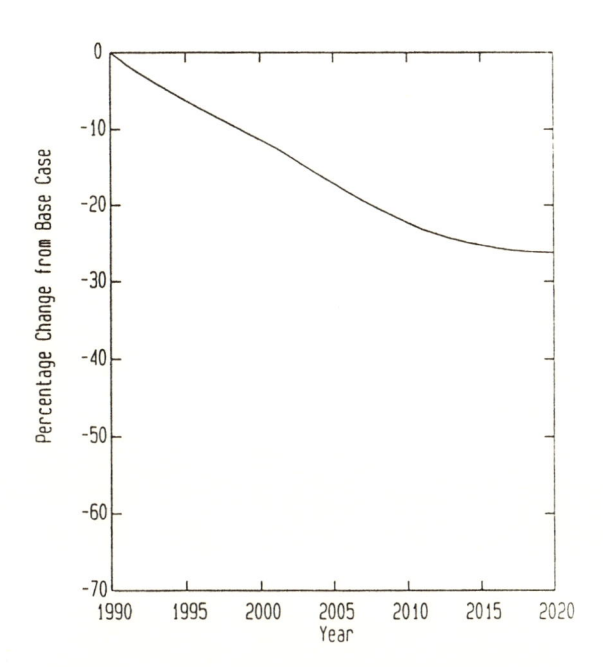

Figure 7. Coal Production Under a Carbon Tax

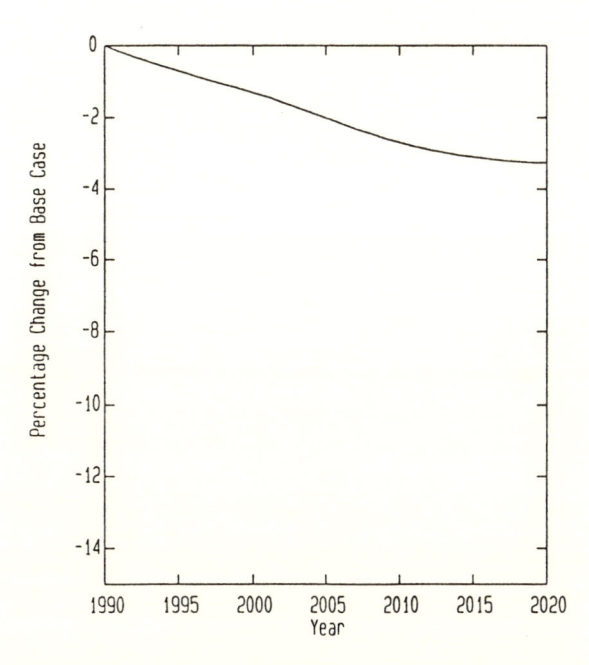

Figure 8. Crude Oil Extraction Under a Carbon Tax

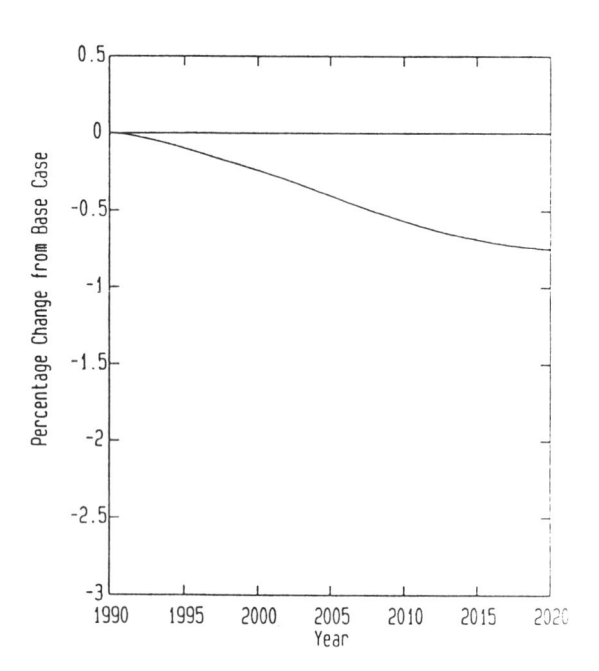

Figure 9. Capital Stock Under a Carbon Tax

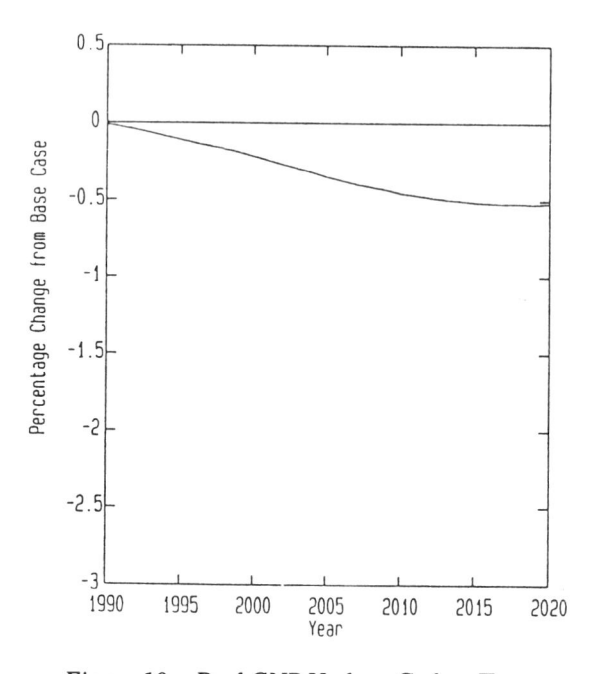

Figure 10. Real GNP Under a Carbon Tax

Reducing U.S. Carbon Dioxide Emissions

its base case level. In fact, under the carbon tax simulation average annual GNP growth over the period 1990–2020 is 0.02 percentage points lower than in the base case.[33] About half of this is due to slowing technical change and half due to slower capital accumulation.

The other two carbon control targets we examined showed dynamic behavior qualitatively similar to that described above. These results can best be displayed by plotting each variable's values for all three simulations on a single graph.[34] Figure 11, for example, shows the paths of the carbon tax needed to achieve each of the targets. The highest path is the tax required to reduce emissions to 20 percent below their 1990 levels; the central path is that for maintaining 1990 emissions; and the lowest path is the tax needed to stabilize emissions at year 2000 levels. Similarly, Figure 12 shows the carbon reductions achieved under each of the policies.[35] Plotting three curves on each figure makes it easy to compare different targets. For example, many of the figures show that as the target becomes more stringent, the variable of interest is pushed further away from the base case. However, some of the figures show much more interesting behavior, and we will focus on these for the remainder of this section.

The first feature to note, which is apparent from Figure 12, is that the three targets require carbon reductions of roughly 8, 14, and 32 percent. (This was also noted in the section on long run results.) Keeping these reductions in mind, Figure 13 is quite interesting because it shows that coal production does not fall in proportion to the drop in emissions. This occurs because it becomes increasingly costly to drive coal production toward zero. Coal users, notably electric utilities, find it increasingly difficult to substitute away from coal as the amount they use of it decreases. This is reflected in Figure 14, which shows that oil extraction costs rise consequently more sharply as regulations become more stringent.

One of the most interesting results of our study is shown in Figure 15, a graph of the capital stock under the three policies. Figure 15 is a very clear example of the effects of intertemporal optimization by households. For the policy which has least effect and occurs furthest in the future (maintaining emissions at year 2000 levels), the early reduction in consumption actually leads to a temporary increase in the capital stock. As explained above, this comes about because households reduce consumption in anticipation of lower future earnings. Only under the most stringent policy (reducing emissions by 20 percent from 1990 levels)

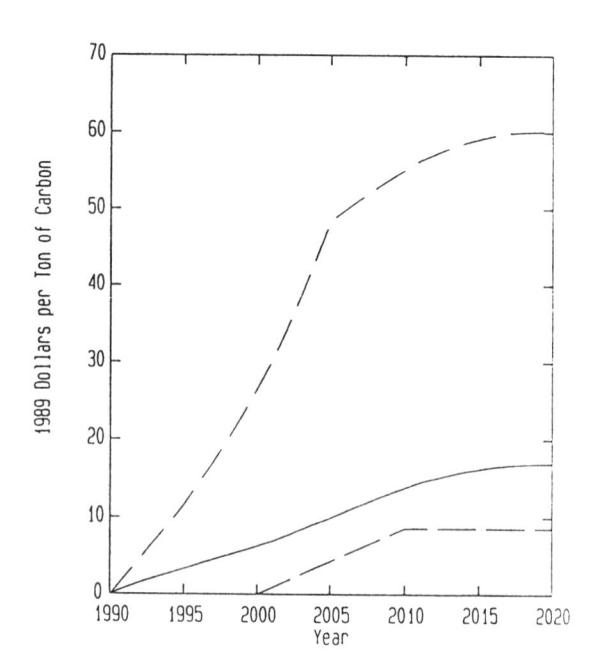

Figure 11. Carbon Taxes for Different Targets

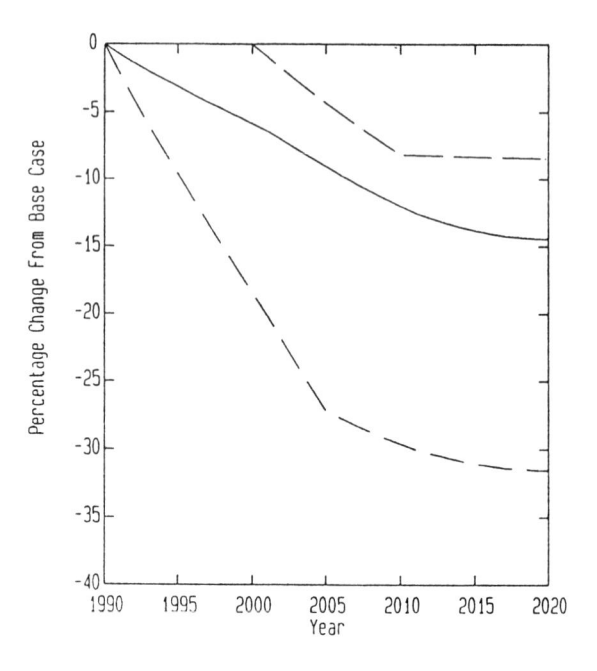

Figure 12. Carbon Emissions Under a Carbon Tax

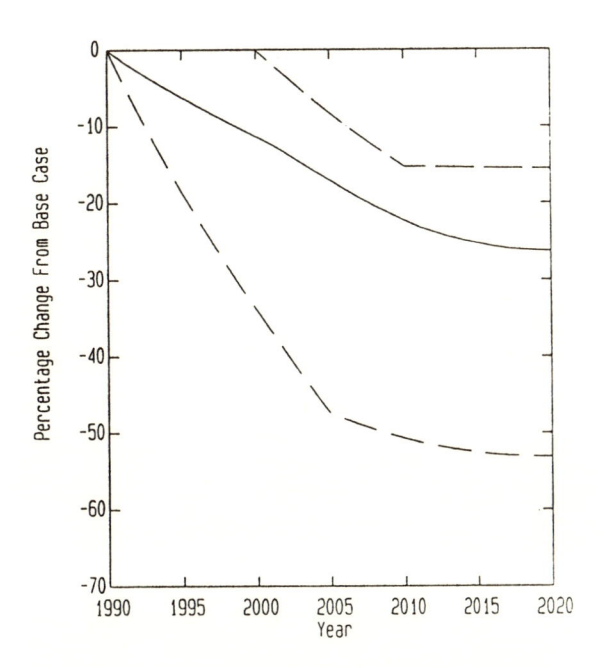

Figure 13. Coal Production Under a Carbon Tax

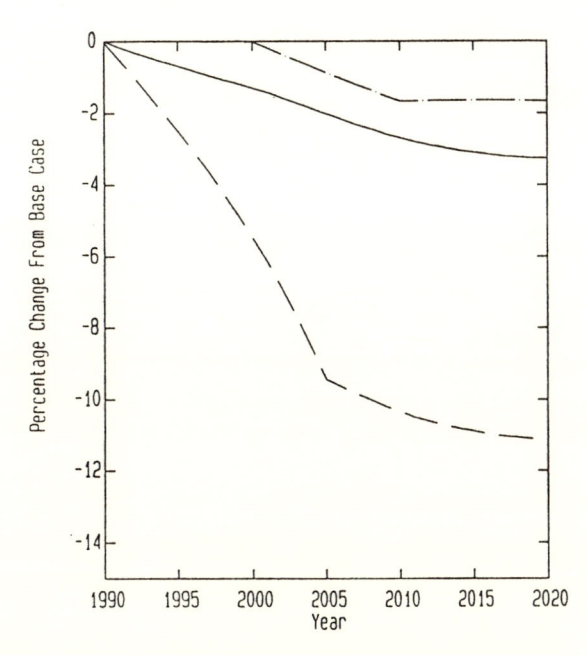

Figure 14. Crude Oil Extraction Under a Carbon Tax

151

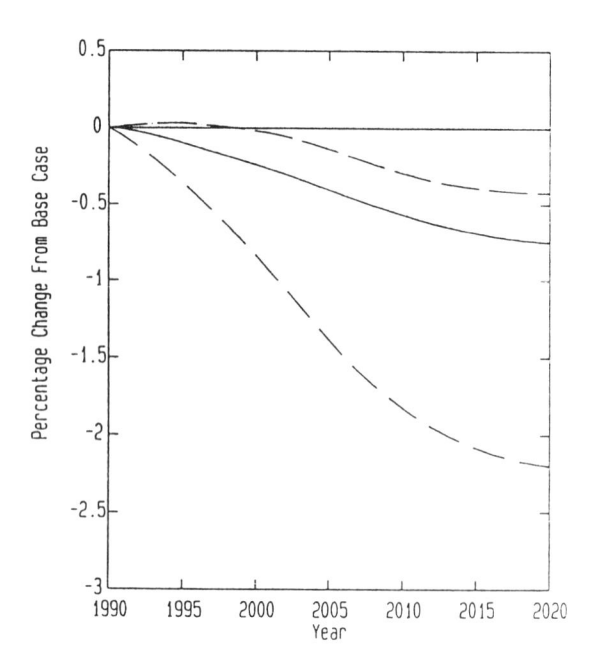

Figure 15. Capital Stock Under a Carbon Tax

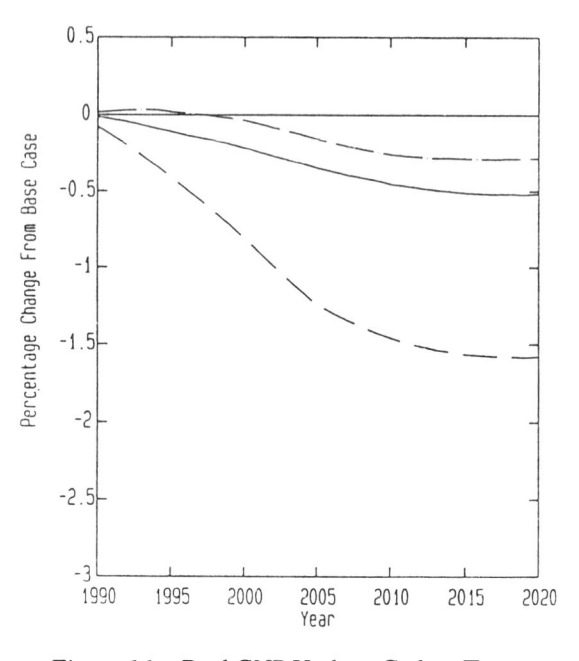

Figure 16. Real GNP Under a Carbon Tax

does the capital stock begin to fall immediately. Finally, the results for GNP, shown in Figure 16, echo those for the capital stock. As mentioned above, GNP falls in part because of the drop in capital accumulation and in part because higher energy prices reduce the rate of technical change.

IV. CONCLUSION

Several important observations can be made about the carbon tax simulations presented in this paper. First, the principal effects of a carbon tax will be felt at the industry level. Coal mining, in particular, will be strongly affected. Even under the least restrictive policy, coal output will fall 16 percent from its base-case value; more restrictive policies could lead to reductions of 50 percent or more. Electric utilities will also be affected, with output falling by as much as 15 percent under tighter emissions restrictions. At more aggregate levels, however, our results also show that the economy-wide effects of a carbon tax would be fairly modest. Thus, the effects of a tax would be concentrated in coal mining, although a handful of other sectors, such as oil extraction, would also be affected.

A second but no less important observation about these simulations is that the stringency and timing of the tax will determine how strongly the tax affects the economy. Comparing long-run results of the three simulations considered here shows that increasingly stringent targets are increasingly costly. Thus, it is essential that any serious national or international goal on carbon dioxide emissions be chosen by carefully comparing costs and benefits. Setting goals arbitrarily risks seriously under- or overemphasizing carbon dioxide reductions. Because our model does not compute benefits, we cannot say which policy is most appropriate. We do urge, however, that careful consideration of benefits be given before any national or international policy is adopted.

ACKNOWLEDGMENTS

This research was sponsored by the Environmental Protection Agency under contract 68–W8–0113. We are grateful for the help and advice of our colleagues Mun Sing Ho, Richard Goettle, Edward Hudson, Barbara Fraumeni, Daekeun Park, Daniel Slesnick, Joel Scheraga, and Michael Shelby. Needless to say, we alone are responsible for any remaining deficiencies.

NOTES

1. The most notable exception is the work of Nordhaus (1991), who has devoted a great deal of effort to measuring the benefits of reducing global warming in order to be able to calculate the optimal reduction.

2. In Jorgenson and Wilcoxen (1991) we also examined the effect of reducing carbon dioxide emissions by imposing taxes on the energy content of primary fuels or by imposing ad valorem fuel taxes.

3. A very thorough discussion of the greenhouse effect and numerous references to the literature are given by EPA (1989).

4. Schneider (1989, p. 774). However, there is a wide spectrum of scientific opinion, as described by Stevens (1989).

5. Schneider (1989, p. 772). This is also subject to dispute, as pointed out by Stevens (1989).

6. Schneider (1989, p. 771). A detailed analysis of the potential economic impact of the legislation introduced by Senator Wirth is presented by the Congressional Budget Office (1990).

7. See, for example, National Resources Defense Council (1989).

8. Many other policies have been suggested; see EPA (1989).

9. The Edmonds-Reilly model has subsequently been used by Edmonds and Reilly (1985), Reilly, Edmonds, Gardner, and Brenkert (1987), the Environmental Protection Agency (1989), the Congressional Budget Office (1990), and others.

10. It is a descendant of Manne's earlier work on ETA-MACRO. See Manne (1981).

11. A nontechnical description of the results of Manne and Richels is presented by Passell (1989).

12. Nordhaus (1991) is one of a series of studies beginning with Nordhaus (1977). Additional references are given by Nordhaus (1979, 1982), Nordhaus and Ausubel (1983), and Nordhaus and Yohe (1983).

13. Sectors 3, 4, 16, 30 and 31 in Table 1.

14. Data on inter-industry transactions are based on input-output tables for the U.S. constructed by the Bureau of Economic Analysis (1984). Income data are from the U.S. national income and product accounts, also developed by the Bureau of Economic Analysis (1986). The data on capital and labor services are discussed by Jorgenson (1990b). Additional details are given by Wilcoxen (1988), Appendix C, and Ho (1989).

15. See Mansur and Whalley (1984) for more detail. A example of the calibration approach is Borges and Goulder (1990).

16. For a more complete discussion of the econometric approach see Jorgenson (1982, 1984).

17. Our approach to endogenous productivity growth was originated by Jorgenson and Fraumeni (1981). The implementation of a general equilibrium model of production that incorporates both substitution among inputs and endogenous productivity growth was discussed by Jorgenson (1984). This model has been analyzed in detail by Hogan and Jorgenson (1991). Further details on the econometric methodology are presented by Jorgenson (1986).

18. The Euler equation approach to modeling intertemporal consumer behavior was originated by Hall (1978). Our application of this approach to full consumption follows Jorgenson and Yun (1986).

19. This allows our model of personal consumption to be used to represent the behavior of individual households, as in Jorgenson and Slesnick (1985). Further details on the econometric methodology are given by Jorgenson (1984, 1990).

20. See Wilcoxen (1988) and Ho (1989) for more details.

21. The relationship between the price of investment goods and the price of capital services is discussed in more detail by Jorgenson (1989).

22. More accurately, between the industries themselves and between industries and final demand categories. Households, in particular, purchase a considerable amount of capital services.

23. Without a model of Congressional decision-making, we must take either the level of government expenditures or the size of the budget deficit to be exogenous.

24. This is the Armington (1969) approach.

25. See Wilcoxen (1988) or Ho (1989) for more details.

26. Unlike ordinary pollutants, carbon dioxide is one of the natural products of combustion. Little can be done to change the amount of it produced when burning any particular fuel.

27. To convert to tons of carbon dioxide, multiply by 3.67.

28. Environmental Protection Agency (1990), internal memorandum.

29. Our breakdown of the U.S. population by age, educational attainment, and sex is based on the system of demographic accounts compiled by Jorgenson and Fraumeni (1989). The population projections are discussed in detail by Wilcoxen (1988, Appendix B).

30. All dollar amounts are in 1989 prices.

31. As noted in Section II, our population forecast is based on work done by the Social Security Administration. Two notable features are that the U.S. population stabilizes early in the next century, and that educational attainment (and hence labor quality) stabilizes as well.

32. Since revenue earned by the tax is given back to households through a vertical shift in the labor tax schedule, the simulation is essentially the replacement of a lump sum tax (the labor tax) by a distorting one (the carbon tax).

33. The difference in two variables growing at rates differing by 0.02 percentage points is about 2 percent after a hundred years.

34. Recall that the targets were (1) maintaining 1990 emissions, (2) reducing emissions by 20 percent below 1990 levels, and (3) gradually introducing taxes to stabilize at year 2000 emissions.

35. Notice that target policies are drawn using the same line type in each graph. Maintaining 1990 emissions is always a solid line, reducing emissions to 20 percent below 1990 is always dashed, and maintaining emissions at 2000 levels is alternating dots and dashes. Also, variables are plotted on the same scale across different tax instruments for easier comparison.

REFERENCES

Armington, P.S. 1969. "The Geographic Pattern of Trade and the Effects of Price Changes." *IMF Staff Papers*, 16, 2:176–199.

Borges, A.M., and L.H. Goulder. 1984. "Decomposing the Impact of Higher Energy Prices on Long-Term Growth." Pp. 319–362 in *Applied General Equilibrium Analysis*, edited by H.E. Scarf and J.B. Shoven. Cambridge: Cambridge University Press.

Bureau of Economic Analysis. 1984. "The Input-Output Structure of the U.S. Economy, 1977." *Survey of Current Business*, 64, 5:42–79.

Bureau of Economic Analysis. 1986. *The National Income and Product Accounts of the United States, 1929–1982: Statistical Tables.* Washington, DC: U.S. Department of Commerce.

Congressional Budget Office. 1990. *Carbon Charges*. Washington, DC: U.S. Government Printing Office.

Department of Energy. 1990. *Annual Energy Review 1990*. Washington, DC: Energy Information Administration.

Edmonds, J.A., and J.M. Reilly. 1983. "Global Energy and CO_2 to the Year 2050." *The Energy Journal*, 4, 3:21–47.

Edmonds, J.A., and J.M. Reilly. 1985. Global Energy—Assessing the Future. New York: Oxford University Press.

Environmental Protection Agency. 1988. *The Potential Effects of Global Climate Change in the United States, Draft Report to Congress*.

Environmental Protection Agency. 1989. *Policy Options for Stabilizing Global Climate, Draft Report to Congress*, 3 Vols.

Environmental Protection Agency. 1990. Internal memorandum.

Hall, R.E. 1978. "Stochastic Implications of the Life Cycle-Permanent Income Hypothesis: Theory and Evidence." *Journal of Political Economy*, 86, 6:971–988.

Ho, M.S. 1989. "The Effects of External Linkages on U.S. Economic Growth: A Dynamic General Equilibrium Analysis." Ph.D Dissertation, Harvard University.

Hogan, W.W., and D.W. Jorgenson. 1991. "Productivity Trends and the Costs of Reducing Carbon Dioxide Emissions." *The Energy Journal*, 12, 1:67–85.

Houghton, R.A., and G.M. Woodwell. 1989. "Global Climate Change." *Scientific American*, 260, 4.

Jorgenson, D.W. 1982. "Econometric and Process Analysis Models for the Analysis of Energy Policy." Pp. 9–62 in *Perspectives in Resource Policy Modeling: Energy and Minerals,* edited by R. Amit and M. Auriel. Cambridge, MA: Ballinger.

Jorgenson, D.W. 1984. "Econometric Methods for Applied General Equilibrium Analysis. Pp. 139–203 in *Applied General Equilibrium Analysis*, edited by E. Scarf and J.B. Shoven. Cambridge: Cambridge University Press.

Jorgenson, D.W. 1986. "Econometric Methods for Modelling Producer Behavior," edited by Z. Griliches and M.D. Intriligator. Pp. 1842–1915 in *Handbook of Econometrics*, Vol. 3, Amsterdam, North-Holland.

Jorgenson, D.W. 1989. "Capital as a Factor of Production." Pp. 1–36 in *Technology and Capital Formation*, edited by D.W. Jorgenson and R. Landau. Cambridge, MA: MIT Press.

Jorgenson, D.W. 1990a. "Aggregate Consumer Behavior and the Measurement of Social Welfare." *Econometrica*, 58, 3.
Jorgenson, D.W. 1990b. "Productivity and Economic Growth." Pp. 19–118 in *Fifty Years of Economic Measurement*, edited by E. Berndt and J. Triplett. Chicago: University of Chicago Press.
Jorgenson, D.W., and B.M. Fraumeni. 1981. "Relative Prices and Technical Change." Pp. 17–47 in *Modeling and Measuring Natural Resource Substitution*, edited by E. Berndt and B. Field. Cambridge, MA: MIT Press.
Jorgenson, D.W., and B.M. Fraumeni. 1989. "The Accumulation of Human and Non-human Capital, 1948–1984." In *The Measurement of Saving, Investment, and Wealth*, edited by R.E. Lipsey and H.S. Tice. Chicago: University of Chicago Press, forthcoming.
Jorgenson, D.W., and D.T. Slesnick. 1985. "General Equilibrium Analysis of Economic Policy," Pp. 293–370 in *New Developments in Applied General Equilibrium Analysis*, edited by J. Piggott and J. Whalley. Cambridge: Cambridge University Press.
Jorgenson, D.W., and D.T. Slesnick. 1987. "Aggregate Consumer Behavior and Household Equivalence Scales." *Journal of Business and Economic Statistics*, 5, 2: 219–232.
Jorgenson, D.W., and K.-Y. Yun. 1986. "The Efficiency of Capital Accumulation." *Scandinavian Journal of Economics*, 88, 1:85–107.
Jorgenson, D.W., and P.J. Wilcoxen. 1990a. "Environmental Regulation and U.S. Economic Growth." *The Rand Journal of Economics,* 21, 2:314–340.
Jorgenson, D.W., and P.J. Wilcoxen. 1990b. "Intertemporal General Equilibrium Modeling of U.S. Environmental Regulation." *Journal of Policy Modeling* (December).
Jorgenson, D.W., and P.J. Wilcoxen. 1991. "Reducing U.S. Carbon Dioxide Emissions: An Assessment of Different Instruments." Unpublished manuscript.
Jorgenson, D.W., L.J. Lau, and T.M. Stoker. 1982. "The Transcendental Logarithmic Model of Aggregate Consumer Behavior." Pp. 9–238 in *Advances in Econometrics*, Vol. 1, edited by R.L. Basmann and G. Rhodes. Greenwich, CT: JAI Press.
Manne, Alan S. 1981. "ETA-MACRO: A User's Guide." Unpublished manuscript, Electric Power Research Institute, Palo Alto, CA.
Manne, A.S., and R.C. Richels. 1990. "CO_2 Emission Limits: An Economic Analysis for the USA." *The Energy Journal,* 11, 2: 51–85.
Mansur, A., and J. Whalley. 1984. "Numerical Specification of Applied General Equilibrium Models: Estimation, Calibration, and Data." Pp. 69–127 in *Applied General Equilibrium Analysis*, edited by H.E. Scarf and J.B. Shoven. Cambridge: Cambridge University Press.
National Research Council. 1983. *Changing Climate.* Washington, DC: National Academy Press.
Natural Resources Defense Council. 1989. "Cooling the Greenhouse: Vital First Steps to Combat Global Warming." Mimeo.
Nordhaus, W.D. 1977. "Economic Growth and Climate: The Carbon Dioxide Problem." *American Economic Review,* 67, 1: 341–346.
Nordhaus, W.D. 1979. *The Efficient Use of Energy Resources.* New Haven, CT: Yale University Press.

Nordhaus, W.D. 1982. "How Fast Should We Graze the Global Commons?" *American Economic Review*, 72, 2: 242–246.

Nordhaus, W.D. 1989. "A Sketch of the the Economics of the Greenhouse Effect." *American Economic Review*, 81, 2: 146–150.

Nordhaus, W.D., and J. Ausubel. 1983. "A Review of Estimates of Future Carbon Dioxide Emissions." In *Changing Climate*. Washington, DC: National Academy Press.

Nordhaus, W.D., and G. Yohe. 1983. "Future Carbon Dioxide Emissions from Fossil Fuels." In *Changing Climate*, National Research Council. Washington, DC: National Academy Press.

Passell, P. 1989. "Staggering Cost Is Foreseen to Curb Warming of Earth." *New York Times* (November 19).

Reilly, J.M., J.A. Edmonds, K.H. Gardner, and A.L. Brenker. 1987. "Uncertainty Analysis of the IEA/ORAU CO_2 Emissions Model." *The Energy Journal*, 8, 3: 1–29.

Schneider, S.H. 1989. "The Greenhouse Effect: Science and Policy." *Science*, 243, 4894: 771–781.

Stevens, W.K. 1989. "Skeptics Challenging Dire 'Greenhouse' Views." *New York Times* (December 13).

Whalley, J., and R. Wigle. 1990. "Cutting CO_2 Emissions: The Effects of Alternative Policy Approaches." Unpublished manuscript.

Wilcoxen, P.J. 1988. "The Effects of Environmental Regulation and Energy Prices on U.S. Economic Performance." Ph.D Dissertation, Harvard University.

DYNAMIC MODELS OF INPUT DEMANDS:
A COMPARISON UNDER DIFFERENT FORMULATIONS OF ADJUSTMENT COSTS

G. C. Watkins and Ernst R. Berndt

I. INTRODUCTION

In their survey article on dynamic factor demands, Berndt, Morrison, and Watkins (1981) classified dynamic models of input demands into first, second and third generation. First generation models are essentially of an ad hoc single equation variety and do not incorporate economic optimization. Second generation models recognize the interrelatedness of factor "disequilibria," but do not explicitly incorporate the movement of quasi-fixed inputs towards equilibrium values within a dynamic optimization framework, and do not identify costs of adjustment associated with quasi-fixed inputs. By contrast, both the path and the costs of adjustment over time of quasi-fixed inputs to equilibrium values are captured by third generation models.

Most empirical applications of third generation models have assumed that adjustment costs of the "internal" variety are a function of net rather than gross investment.[1] An implicit assumption in such treatment is that replacement investment is "frictionless." Pindyck and Rotemberg (1983) have suggested this holds for "external" adjustment costs. Morrison (1982) has shown that internal adjustment costs are of greater significance than the external variety, but it is not clear that treating internal adjustment costs as a function of net rather than gross investment is legitimate.

This paper examines alternative third generation dynamic models, allowing internal adjustment costs to be a function of either net or gross investment, and assesses whether such differences are empirically important. We undertake this task by estimating models relating to the Canadian manufacturing sector over the period 1957–1982. We find that significant and systematic differences in model results do emerge, suggesting that the question of the appropriate formulation of adjustment costs in such models is empirically significant.

The paper consists of two main sections. Section II deals with differences in model specification between the respective "gross" and "net" investment adjustment cost formulations. Section III concerns empirical estimates of both model structure and of factor elasticities derived for three sets of Canadian data: manufacturing in aggregate, textiles, and iron and steel. Our conclusions are summarized in Section IV.

II. MODEL SPECIFICATION

We deal first with specification of the "net investment" adjustment cost version of a third generation dynamic model. Since this model has been outlined several times in the literature (Berndt, Morrison, and Watkins 1981, pp. 272–279) our treatment of it here is relatively brief. We then show how the model changes when adjustment costs are defined as a function of gross rather than net investment. Finally, we define the models in a form suitable for empirical implementation.

A. Net Investment Version of Adjustment Costs

Define the production function of the firm as

$$Q(t) = F(v(t), x(t), \dot{x}(t), t) \tag{1}$$

A Comparison of Dynamic Models of Input Demands

where $Q(t)$ is gross output representing various efficient combinations of variable inputs (v), quasi-fixed inputs (x), accumulation (or decumulation) in quasi-fixed inputs ($\dot{x}$) and technological change (t). If levels of quasi-fixed inputs vary ($\dot{x} \neq 0$), output falls because of the need to devote efforts to implement such changes rather than to producing output. This diminution in output ($\partial F(t)/\partial \dot{x}(t) < 0$) constitutes internal costs of adjustment.

If we designate the cost of varying quasi-fixed inputs as $c_i(\dot{x}_i)$, then the adjustment cost function at time t should have the properties (i) $c_i(0) = 0$, (ii) $c_i(\dot{x}_i) > 0$, (iii) $c_i'(\dot{x}_i) > 0$, and (iv) $c_i''(\dot{x}_i) > 0$. Condition (i) says that when net investment is zero, there are no adjustment costs; note that this implies replacement investment does not generate internal costs of adjustment. Condition (ii) simply says net investment always involves positive adjustment costs. Condition (iii) says adjustment costs increase as net investment increases, and condition (iv) says adjustment costs rise at an increasing rate for each increment in new investment *within* period t.[3] Thus, these conditions imply positive, increasing marginal internal costs of adjustment. They also assume adjustment costs associated with a given investment are not incurred beyond the year in which the investment takes place.

Firms can be viewed as minimizing normalized variable costs

$$G = \sum_{j=1}^{m} w_j v_j$$

conditional on w_j, Q, x_i, $\dot{x}_i$ and where w_j = normalized price of input j. Under traditional regularity conditions, the normalized restricted cost function

$$G(t) = G(w, x, \dot{x}, Q, t) \qquad (2)$$

can be shown to be increasing and concave in w, increasing and convex in $\dot{x}$, and decreasing and convex in x.

By the Shephard–Uzawa–McFadden lemma, the partial derivative of G with respect to the normalized price of any variable input w_j equals the short-run cost minimizing demand for v_j:

$$\partial G/\partial w_j = v_j \qquad (3)$$

In full equilibrium, the partial derivative of G with respect to the quantity of any quasi-fixed input equals the negative of the normalized rental price of the quasi-fixed input:

$$\partial G/\partial x_i = -u_i \qquad (4)$$

where $u_i = q_i(r + d_i)$ and q_i is the normalized acquisition price of the quasi-fixed input, r is the (after tax) rate of return, and d_i is the rate of depreciation.

By inserting the optimal values for v_j from Equation (3) in Equation (2), $G(t)$ becomes the minimum variable cost obtainable at time t conditional on normalized variable input prices, the level of the quasi-fixed factors $x(t)$, the change in the quasi-fixed factors $\dot{x}(t)$, output $Q(t)$, and the passage of time, t. Therefore, $G(t)$ is a normalized restricted cost function (NRCF), conceptually similar to the normalized restricted profit function discussed by Lau (1976).

The short-run optimization behavior embodied in the normalized restricted cost function yields optimal variable costs conditional on given levels of output $Q(t)$, quasi-fixed inputs $x(t)$ and net investment $\dot{x}(t)$. But the firm would seek to minimize the present value of total costs over time by installing a suitable stock of quasi-fixed inputs (x_i). The appropriate objective function, $L(0)$, can be written:

$$L(0) = \int_0^\infty e^{-rt} \left(\sum_{j=1}^m w_j v_j(t) + \sum_{i=1}^n q_i z_i(t) \right) dt \qquad (5)$$

where r is the firm's real after tax discount rate (subject to static expectations) and z_i is the gross addition to the stock of the ith quasi-fixed factor $(z_i = \dot{x}_i + d_i x_i)$.

The firm's problem is to minimize $L(0)$ with respect to the production function, Equation (1). The solution to this calculus of variations problem is obtained by choosing the time paths $v(t)$, $x(t)$ so as to minimize $L(0)$, given initial stocks of quasi-fixed factors $x(0)$, and with $v(t)$, $x(t) > 0$.

A solution to Equation (5) can be obtained from the Euler-Legendre conditions, assuming static expectations with respect to normalized factor prices and output. Treadway (1974) has shown this can be written as

$$G_{\dot{x}\dot{x}}\ddot{x} + G_{x\dot{x}}\dot{x} - G_x - rG_{\dot{x}} - u = 0 \qquad (6)$$

where the x, $\dot{x}$ subscripts denote derivatives and $\ddot{x}\dot{x}$ is the second partial derivative with respect to time.

A steady state solution x^* which defines the long-run optimal stock of

A Comparison of Dynamic Models of Input Demands

the quasi-fixed factors entails setting $G_{\dot{x}\dot{x}}$ and $G_{x\dot{x}}$ to zero, since at the steady state net investment, $\dot{x}$, and the rate of change of net investment, $\ddot{x}$, will be zero. Thus Equation (6) becomes

$$-G_x(w, x^*) = u + rG_{\dot{x}}(w, x^*) \tag{7}$$

This expression can be interpreted as follows: the left-hand side is the marginal benefit to the firm of changing quasi-fixed inputs (for example, the reduction in variable costs brought about by purchasing more energy-efficient capital equipment), while the right-hand side is the marginal cost (user cost plus the amortized marginal adjustment cost) of a change in the amount of capital services at $\dot{x} = 0$. In equilibrium, marginal benefits must equal marginal costs.

If adjustment costs were treated as a function of net rather than gross changes in the quasi-fixed inputs, it is reasonable to assume $G_{\dot{x}}(w, x^*) = 0$, in which case Equation (7) would be identical to the steady state marginal condition obtained when adjustment costs were not explicitly considered.

The internal cost of adjustment model yields clearly defined short-run variable input demand Equation (3) and is based on explicit dynamic optimization. Treadway (1974) has linked this type of model to the partial adjustment literature by showing that $\dot{x}$ can be generated from Equation (7) as an approximate solution (in the neighborhood of $x^*(t)$) to the multivariate linear differential equation system

$$\dot{x} = M^*(x^* - x) \tag{8}$$

assuming the cost function were globally quadratic.[4] The adjustment matrix, M^*, is determined from the solution to the quadratic form:

$$G^*_{\dot{x}\dot{x}}M^{*2} + rG^*_{\dot{x}\dot{x}}M^* - (G^*_{xx} + rG^*_{x\dot{x}}) = 0 \tag{9}$$

In the case of a single quasi-fixed factor, x_1, the optimal net capital accumulation equation is

$$\dot{x}_1 = M^*_1(x^*_1 - x_1) \tag{10}$$

Solving (9) for the positive root yields the expression for the adjustment parameter,

$$M^*_1 = -\frac{1}{2}\left\{ r - \left[r^2 + \frac{4(G^*_{x_1 x_1} + rG^*_{x_1 \dot{x}_1})}{G^*_{\dot{x}_1 \dot{x}_1}} \right]^{1/2} \right\} \tag{11}$$

and at the stationary point in this formulation, by assumption $G_{xx}^* = 0$.

B. Gross Investment Version of Adjustment Costs

How does the formulation of the model change when adjustment costs depend on gross rather than net investment? As becomes apparent below, inclusion of gross investment in the adjustment process leads to a more complex model.

We want to make the production function (1) dependent on gross (z) rather than net investment ($\dot{x}$). So we write

$$Q(t) = F[v(t), x(t), z(t), t] \qquad (12)$$

where $z(t) =$ gross investment, $[z_i]$, $i = 1, \ldots, n$; $z_i = \dot{x}_i + d_i x_i$; and $d_i =$ depreciation rate, quasi-fixed factor i. Internal costs of adjustment now constitute $\partial(t)/\partial z(t) < 0$.

The normalized restricted cost function, G, becomes

$$G(t) = G(w, x, z, Q, t) \qquad (13)$$

and is concave in w, increasing and convex in z and decreasing and convex in x. These conditions parallel those under the net investment formulation, but the condition $G_z > 0$ requires comment. The supposition is that variable costs will rise in the current period as gross investment increases, due to internal adjustment costs. In the next period, when such adjustment costs are no longer present, variable costs normally will fall by virtue of the increase in the stock of quasi-fixed inputs associated with gross investment in the previous period ($G_x < 0$).

The expression (5) for the present value of a future stream of production costs, given output, which the firm seeks to minimize remains the same, but the underlying production function (12) differs from (1) in that gross rather than net investment affects the amount and composition of inputs used to generate output.

The development of the model proceeds exactly as beforehand, except that the restricted cost function G reflects gross rather than net investment. In particular, Equation (7) can be written as before. Its interpretation was that the marginal benefit to the firm of changing a quasi-fixed input equates to its user cost plus the amortized marginal adjustment cost. However, in the steady state the adjustment costs for quasi-fixed inputs now become a function of replacement investment, whereas in the net investment formulation adjustment costs in the steady state were zero.

A Comparison of Dynamic Models of Input Demands

In other words, in the net investment formulation *there is an implicit assumption that replacement investment is frictionless.* Here we drop this assumption.

At the steady state equilibrium, gross investment is simply

$$z_i = d_i x_i \tag{14}$$

since net investment, $\dot{x}$, is zero. The direct (normalized) cost of gross investment, now just replacement investment, is $q_i z_i = q_i d_i x_i$.

As before, the internal cost of adjustment model outlined above can be linked to the "partial adjustment" literature by showing that $\dot{x}$ can be generated as an approximate solution to a multivariate linear differential equation system.

C. Empirical Implementation

The empirical implementation employs a four input production function—capital, labor, energy, and intermediate materials—where capital is treated as a quasi-fixed factor, and energy, labor, and intermediate materials are variable inputs.[5]

If input and production prices were exogenously determined and capital were fixed in the short run, the theory of duality between cost and production functions discussed beforehand implies that, given short-run cost-minimizing behavior, the underlying production function can be uniquely represented by the normalized restricted variable cost function

$$G(P_E, P_L, P_M, K_{-1}, \Delta K + dDK_{-1}, Q, t) = L + P_E E + P_M M \tag{15}$$

where P_E = price of energy
 P_L = price of labor
 P_M = price of non-energy intermediate materials
 K = capital stock
 ΔK = change in capital stock, approximated empirically by first differences
 d = depreciation rate
 D = dummy variable, 0 or 1
 Q = gross output
 t = time, intended to represent disembodied technical change

The RHS of this expression is simply total (normalized) variable cost, where the numeraire is the price of labor. The functional form chosen to approximate the normalized restricted cost function is quadratic.[6]

The general formulation (15) of the normalized restricted cost function covers both the case where adjustment costs are treated as a function of gross investment ($D = 1$) or net investment ($D = 0$).

In models involving specification of technical change and without the imposition of constant long-run returns to scale, often the empirical results show strong increasing long-run returns to scale (LRCRTS) accompanied by technological regression. In other words, separation of the scale and technology effects tends to be difficult. There is a fair degree of empirical evidence for LRCRTS (see Jorgensen, 1972; Lucas, 1967), especially for aggregative industrial sector studies whereby although at the firm level increasing returns to scale may hold, at the industry level replication of firm size makes LRCRTS plausible.

Accordingly, the model is specified in a form that imposes strict LRCRTS, at least for the net investment version. It also imposes certain symmetry conditions on the coefficients, as required by symmetry of second derivatives.

The normalized restricted variable cost function represented by a quadratic approximation is

$$G = L + P_E E + P_M M = Q[a_0 + a_{0t}t + a_E P_E + a_M P_M + a_{Mt} P_M t$$

$$+ a_{Et} P_E t + 1/2(a_{EE} P_E^2 + a_{MM} P_M^2) + a_{EM} P_E P_M]$$

$$+ 1/2 a_{KK} K_{-1}^2/Q + a_K K_{-1} + a_{Kt} K_{-1} t + a_{EK} P_E K_{-1} + a_{MK} P_M K_{-1}$$

$$+ a_{\dot{K}t}(\Delta K + dDK_{-1})t + a_{\dot{K}}(\Delta K + dDK_{-1})$$

$$+ a_{E\dot{K}} P_E(\Delta K + dDK_{-1}) + a_{M\dot{K}} P_M(\Delta K + dDK_{-1})$$

$$+ a_{K\dot{K}} K_{-1}(\Delta K + dDK_{-1})/Q + 1/2 a_{\dot{K}\dot{K}}(\Delta K + dDK_{-1})^2/Q \qquad (16)$$

1. Net Investment Version of Adjustment Costs ($D = 0$)

Internal costs of adjustment $C(\Delta K)$ are represented by all the terms of variable cost function G given by Equation (16) involving K:

$$C(\Delta K) = a_{\dot{K}} \Delta K + a_{\dot{K}t} \Delta K t + a_{E\dot{K}} P_E \Delta K + a_{M\dot{K}} P_M \Delta K + a_{K\dot{K}} K_{-1} \Delta K/Q$$

$$+ 1/2 a_{\dot{K}\dot{K}} \Delta K^2/Q$$

At the stationary optimal point defined by the fully adjusted capital stock,

A Comparison of Dynamic Models of Input Demands

K^*, ΔK must equal zero, and correspondingly so must marginal adjustment costs:

$$C'(\Delta K)_{\Delta K=0} = a_{\dot{K}} + a_{\dot{K}t}t + a_{E\dot{K}}P_E + a_{M\dot{K}}P_M + a_{K\dot{K}}K^*/Q = 0$$

Satisfaction of this condition requires imposition of the restrictions

$$a_{\dot{K}} = a_{\dot{K}t} = a_{E\dot{K}} = a_{M\dot{K}} = a_{K\dot{K}} = 0 \tag{17}$$

Using the conditional variable cost optimizing relation (2), from (16) we obtain short-run optimal input/output demand equations for energy (E) and materials (M) as

$$E/Q = a_E + a_{Et}t + a_{EE}P_E + a_{EM}P_M + a_{EK}K_{-1}/Q \tag{18}$$

$$M/Q = a_M + a_{Mt}t + a_{MM}P_M + a_{EM}P_E + a_{MK}K_{-1}/Q \tag{19}$$

The short-run demand equation for labor is derived as a residual from Equation (16): $L/Q = G/Q - P_E E/Q - P_M M/Q$,

$$L/Q = a_0 + a_{0t}t - 1/2(a_{EE}P_E^2 + 2a_{EM}P_M P_E + a_{MM}P_M^2) + a_K K_{-1}/Q$$

$$+ a_{Kt}K_{-1}t/Q + 1/2 a_{KK}K_{-1}^2/Q^2 + 1/2 a_{\dot{K}\dot{K}}\Delta K^2/Q^2 \tag{20}$$

If we differentiate (16) with respect to K, impose (17), and, given $\dot{K} = 0$ in equilibrium, we have

$$G_K^* = a_{KK}K^*/Q + a_K + a_{Kt}t + a_{EK}P_E + a_{MK}P_M \tag{21}$$

The earlier equilibrium conditions entail

$$-G_K^* - rG_{\dot{K}}^* - u_K = 0 \tag{22}$$

where u_K is the (normalized) user cost of capital. At $\Delta K = 0$, $G_{\dot{K}}^* = 0$, so solving (22) for the equilibrium capital stock, K^*, yields

$$K^* = -(a_K + a_{Kt}t + a_{EK}P_E + a_{MK}P_M + u_K)(Q/a_{KK}) \tag{23}$$

Also, we have

$$\Delta K = K - K_{-1} = M^*(K^* - K_{-1}) \tag{24}$$

The general expression for M^* is given by (11), while from (16) $G_{KK} = a_{KK}/Q$, $G_{\dot{K}\dot{K}} = a_{\dot{K}\dot{K}}/Q$, and $G_{K\dot{K}}/Q = a_{K\dot{K}}/Q$. Thus, after substituting (11) and imposing $a_{K\dot{K}} = 0$ from constraints (17), we obtain

$$M^* = -1/2\left[r - \left[r^2 + \frac{4a_{KK}}{a_{\dot{K}\dot{K}}}\right]^{1/2}\right] \tag{25}$$

Substituting Equation (25) and (23) in Equation (24) yields

$$\Delta K/Q = -1/2 \left\{ r - \left[r^2 + \frac{4a_{KK}}{a_{\dot{K}\dot{K}}} \right]^{1/2} \right\}$$
$$\left[\frac{1}{a_{KK}} (a_K + a_{K_l}t + a_{EK}P_E + a_{MK}P_M + u_K) + K_{-1}/Q \right] \quad (26)$$

The equation system, then, for the net investment third generation model with LRCRTS imposed consist of expressions (18), (19), (20), and (26), the expressions for energy, materials, labor, and net investment, respectively.

2. Gross Investment Version of Adjustment Costs $(D = 1)$[7]

The internal costs of adjustment (C) are represented by all the terms relating to gross investment $I_g = \Delta K + dK_{-1}$ from Equation (16). Thus, we have

$$C(z) = [a_{\dot{K}} + 1/2 a_{\dot{K}\dot{K}} \Delta K/Q + a_{E\dot{K}}P_E + a_{M\dot{K}}P_M + a_{\dot{K}\dot{K}}K_{-1}/Q$$
$$+ a_{\dot{K}t}t] \Delta K + [a_{\dot{K}} + 1/2 a_{\dot{K}\dot{K}} dK_{-1}/Q + a_{\dot{K}\dot{K}} \Delta K/Q + a_{E\dot{K}}P_E \quad (27)$$
$$+ a_{M\dot{K}}]P_M + a_{\dot{K}\dot{K}}K_{-1}/Q + a_{\dot{K}t}t]dK_{-1}$$

At the optimal stationary point, $\Delta K = 0$ and costs of adjustment reduce to

$$C(z)_{\Delta K=0, K=K^*} = [a_{\dot{K}} + 1/2 a_{\dot{K}\dot{K}} dK^*/Q + a_{E\dot{K}}P_E + a_{M\dot{K}}P_M + a_{\dot{K}\dot{K}}K^*/Q$$
$$+ a_{\dot{K}t}t]dK^* \quad (28)$$

From (27) *marginal* internal costs of adjustment with respect to gross (or net) investment are

$$C'(z) = a_{\dot{K}} + a_{\dot{K}\dot{K}}I_g/Q + a_{E\dot{K}}P_E + a_{M\dot{K}}P_M + a_{\dot{K}\dot{K}}K_{-1}/Q + a_{\dot{K}t}t \quad (29)$$

and for these marginal costs of adjustment to be increasing it is necessary that

$$C''(z) = a_{\dot{K}\dot{K}} > 0 \quad (30)$$

We expect (27), (28) and (29) would all be positive even at long run equilibrium, since adjustment costs would be incurred by installing replacement investment, dK^*.

At a long-run or steady state of equilibrium, internal costs of adjustment relating to *net* investment remain zero. In the context of a net

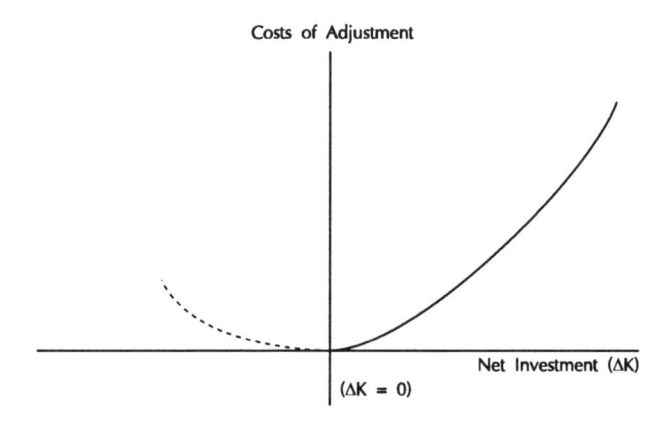

A Comparison of Dynamic Models of Input Demands

Figure 1. Costs of Adjustment and Net Investment Specification

investment formulation of internal costs of adjustment, then, the tangent to the cost of adjustment function at $\Delta K = 0$ must itself have zero slope (see Figure 1). This of course treats net investment as the only variable affecting adjustment costs. If under this supposition marginal adjustment costs were positive at $\Delta K = 0$, it would be economic to disinvest, and if marginal adjustment costs were negative at $\Delta K = 0$, it would be economic to increase the capital stock, K. Thus, only if $C'(\Delta K)_{\Delta K=0} = 0$ would the situation be one of steady-state equilibrium. Hence, in the former net investment specification, the condition $C'(\Delta K)_{\Delta K=0} = 0$ was imposed and implied the parameter restrictions (17). There is also a presumption here that disinvestment ($\Delta K < 0$) would involve adjustment costs, since it would increasingly dislocate the production process. This is indicated by the dotted curve in the north-west quadrant of Figure 1.

In the *gross* investment specification of internal costs of adjustment (see Figure 2), costs of adjustment would be minimized when gross investment was zero, but this could not correspond to a steady state equilibrium because of course it would imply continual *net* disinvestment. Again, by definition steady state equilibrium can occur only when net investment is zero. At the same time, gross investment will be constrained to equal replacement investment $dK^* = dK_{-1}$. In Figure 2, at steady state equilibrium, gross investment of dK^* is shown to be entailing positive costs of adjustment of an amount A. Note that at point C corresponding to gross investment of dK^*, *marginal* internal costs of adjustment are shown as positive, not zero as they were in the net investment case when $\Delta K = 0$ (see Figure 1).

Hence, a priori it does not seem to make sense to impose the earlier

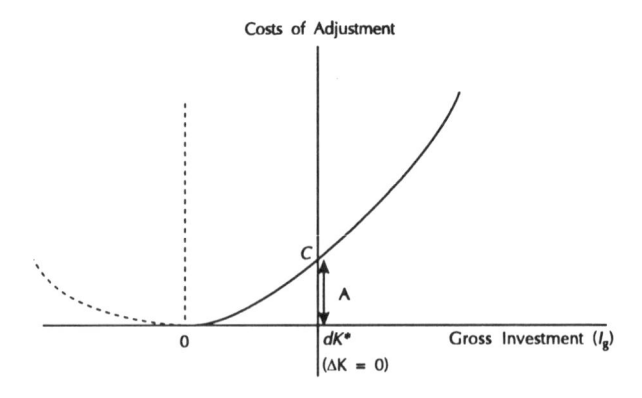

Figure 2. Costs of Adjustment and Gross Investment Specification

restrictions [Equation (17)] in the gross investment model, for if they did hold the terms relating to dK_{-1} in (27) would be zero, and so the entire expression for $C(z)$ given in (27) would be zero when $\Delta K = 0$. This would contradict the supposition that adjustment costs would still be incurred via replacement investment.

We now turn to the derivation of estimating equations for the various inputs. Using relation (2), from (16) and recalling that now $D = 1$, we obtain short run optimal input/output demand equations for energy (E) and materials (M) as

$$E/Q = a_E + a_{Et}t + a_{EE}P_E + a_{EM}P_M + a_{EK}K_{-1}/Q + a_{E\dot{K}}I_g/Q \quad (31)$$

$$M/Q = a_M + a_{Mt}t + a_{MM}P_M + a_{EM}P_E + a_{MK}K_{-1}/Q + a_{M\dot{K}}I_g/Q \quad (32)$$

The difference between these equations and the corresponding ones beforehand [(18) and (19)] is the inclusion of the terms relating to investment. In other words, the cost-minimizing variables E and M input quantities are directly affected by the current rate of investment.

The short-run labor demand to output equation is derived as a residual from Equation (16): $L/Q = G/Q - P_E E/Q - P_M M/Q$, where E/Q and M/Q are given by Equations (31) and (32) respectively. That is,

$$L/Q = a_0 + a_{0t}t - 1/2(a_{EE}P_E^2 + 2a_{EM}P_E P_M + a_{MM}P_M^2)$$

$$+ 1/2a_{KK}K_{-1}^2/Q^2 + a_K K_{-1}/Q + a_{Kt}K_{-1}t/Q + a_{\dot{K}t}I_g t/Q$$

$$+ a_{\dot{K}}I_g/Q + 1/2a_{\dot{K}\dot{K}}I_g^2/Q^2 + a_{K\dot{K}}K_{-1}I_g/Q^2 \quad (33)$$

A Comparison of Dynamic Models of Input Demands

We now derive K*, the long run steady state equilibrium level of capital. From Equation (7), we can write the long-run equilibrium condition as

$$-G^*_K(w, K^*) = u_K + rG^*_{\dot{K}}(w, K^*) \tag{34}$$

From (16), we obtain for the point of evaluation $K = K_{-1}$:

$$\frac{\partial G}{\partial K_t} = G_K = a_{KK}K_{-1}/Q + a_K + a_{K_t}t + a_{EK}P_E + a_{MK}P_M + a_{\dot{K}_t}dt$$

$$+ a_{\dot{K}}d + a_{E\dot{K}}P_Ed + a_{M\dot{K}}P_Md + a_{\dot{K}\dot{K}}dI_g/Q + a_{K\dot{K}}I_g/Q + \frac{a_{KK}dK_{-1}}{Q} \tag{35}$$

When the derivative is evaluated at the optimal capital stock K^* ($K^* = K_{-1}$), $\Delta K = 0$ and thus investment, I_{gt}, is confined to replacement investment dK^*, the investment needed to sustain the optimal capital stock. At this point, the last two terms of (35) can be combined as $2a_{K\dot{K}}dK^*/Q$, while the term attached to $a_{\dot{K}\dot{K}}$ becomes d^2K^*/Q (since $\Delta K = 0$). Thus, we have

$$G^* = a_{KK}K^*/Q + a_K + a_{K_t}t + a_{EK}P_E + a_{MK}P_M + a_{\dot{K}_t}dt + a_{\dot{K}}d$$

$$+ a_{E\dot{K}}P_Ed + a_{M\dot{K}}P_Md + a_{\dot{K}\dot{K}}d^2K^*/Q + 2a_{K\dot{K}}dK^*/Q \tag{36}$$

Also, from (16) we have

$$\frac{\partial G}{\partial \dot{K}} = G_{\dot{K}} = a_{\dot{K}} + a_{\dot{K}_t}t + a_{E\dot{K}}P_E + a_{M\dot{K}}P_M + a_{\dot{K}\dot{K}}\dot{K}/Q + a_{\dot{K}\dot{K}}d_{-1}K_{-1}/Q$$

$$+ a_{K\dot{K}}K_{-1}/Q = a_{\dot{K}} + a_{\dot{K}_t} + a_{E\dot{K}}P_E + a_{M\dot{K}}P_M + a_{\dot{K}\dot{K}}I_{gt}/Q$$

$$+ a_{K\dot{K}}K_{-1}/Q \tag{37}$$

In a steady state $\Delta K = 0$, thus (37) becomes at the point corresponding to K^*:

$$G^*_{\dot{K}} = a_{\dot{K}} + a_{\dot{K}_t}t + a_{E\dot{K}}P_E + a_{M\dot{K}}P_M + a_{\dot{K}\dot{K}}dK^*/Q + a_{K\dot{K}}K^*/Q \tag{38}$$

We can now solve for K*, the long run steady state equilibrium value for capital, by inserting (36) and (38) in (34), yielding

$$K^* = -Q[a_K + a_{K_t}t + a_{EK}P_E + a_{MK}P_M + u_K + (d+r)(a_{\dot{K}_t}t + a_{\dot{K}}$$

$$+ a_{E\dot{K}}P_E + a_{M\dot{K}}P_M]/[a_{KK} + a_{K\dot{K}}d(d+r) + a_{K\dot{K}}(2d+r)] \tag{39}$$

Since we now have an expression for K*, we can also derive an equation for I_g, the rate of gross investment consistent with eventually reaching the equilibrium steady state capital stock.[8]

From (16), we have

$$\frac{\partial^2 G^*}{\partial K^2}\bigg|_{\Delta K=0} = G^*_{KK} = a_{KK}/Q + a_{\dot K \dot K}d^2/Q + 2a_{K\dot K}d/Q \qquad (40)$$

Also, from Equation 37

$$\frac{\partial^2 G^*}{\partial K \partial \dot K}\bigg|_{\Delta K=0} = G^*_{K\dot K} = a_{\dot K \dot K}d/Q + a_{K \dot K}/Q \qquad (41)$$

and from (37) we have

$$\frac{\partial^2 G^*}{\partial \dot K^2}\bigg|_{\Delta K=0} = G^*_{\dot K \dot K} = a_{\dot K \dot K}/Q \qquad (42)$$

Note here again that in equilibrium I_g is simply dK_{-1}.

Thus substituting (4), (41) and (42) in the general expression for the adjustment parameter, M^*, yields

$$M^*_K = -\frac{1}{2}\left\{r - \left[r^2 + \frac{4[a_{KK} + a_{\dot K \dot K}d(d+r) + a_{K \dot K}(2d+r)]}{a_{\dot K \dot K}}\right]^{1/2}\right\} \qquad (43)$$

The investment equation is

$$I_g = B^*_K(K^* - K_{-1}) + dK_{-1} \qquad (44)$$

By substituting (43) and (39) in (44), we obtain,

$$I_g/Q = -\frac{1}{2}\left\{r - \left[r^2 + \frac{4[a_{KK} + a_{\dot K \dot K}d(d+r) + a_{K \dot K}(2d+r)]}{a_{\dot K \dot K}}\right]^{1/2}\right\}$$

$$\{[a_K + a_{K_t}t + a_{EK}P_E + a_{MK}P_M + u_K + (d+r)(a_{\dot K_t}t + a_{\dot K})$$

$$+ a_{E\dot K}P_E + a_{M\dot K}P_M]\} + \{[a_{KK} + a_{\dot K \dot K}d(d+r) + a_{K \dot K}(2d+r)$$

$$+ K_{-1}/Q]\} + dK_{-1}/Q \qquad (45)$$

the endogenous gross investment equation.

We now highlight the differences between the empirical expressions for the net and gross investment versions of "third generation" dynamic models.

The relevant expressions for the net investment model are the cost function (16), the energy demand Equation (18), the materials demand

Equation (19), the labor demand Equation (20), and the endogenous investment Equation (26). Corresponding equations for the gross investment model from this paper are (16), (31), (32), (33) and (45).

Although the cost function for the gross investment model differs from that for net investment simply by replacement of the terms in ΔK by terms in I_g, after imposition of constraints (17) in the net investment version—constraints which are not upheld in the gross investment version—the two cost functions also differ by elimination of the terms $a_{\dot{K}}$, $a_{\dot{K}I}$, $a_{E\dot{K}}$, $a_{M\dot{K}}$, and $a_{K\dot{K}}$ in the net investment model.

The difference in the respective energy and materials equations is that the net investment model does not include a contemporaneous investment term. The impact of current investment is only through the change in the capital stock in the next period. It can be shown (see Watkins, 1987) that this has implications for whether LRCRTS is upheld for the variable inputs. In the net investment version, LRCRTS is satisfied for all inputs, variable or quasi-fixed. In the gross investment version, demands for quasi-fixed inputs are homogenous of degree one in output, but varying rather than constant long run returns to scale may apply to variable inputs. The labor equations—treated as residuals—reflect the respective differences in the cost, energy, and materials functions.

The investment function is a good deal more complex in the gross investment version, reflecting the inclusion of replacement investment. We now turn to an empirical comparison of the two model formulations.

III. EMPIRICAL TESTS

This section examines whether the differences in model specification between the net and gross investment versions are empirically important. We apply both the net and gross investment versions of the models to three sets of data:

1. Canadian Aggregate Manufacturing Sector, 1957–1982;
2. Canadian Iron and Steel Industry, 1957–1982;
3. Canadian Textile Industry, 1957–1982.

Each set of results is discussed below in parallel fashion. In particular, we look at estimates of model coefficients for the net and gross investment specifications of adjustment costs, at variations in estimated elasticities, and at input relationships.

In each case, the respective-model equations cited beforehand are estimated in "input–output" form, where the dependent variable is divided by gross output Q. Note that estimation of the models in the "input–output" form potentially reduces heteroscedasticity. Morrison (1979) has shown that iterated generalized least squares is equivalent to nonlinear full information maximum likelihood estimations procedures for the net investment version equations; Watkins (1984) has shown that the traditional log likelihood procedures also hold for the gross investment version. Accordingly, estimation did not need to rely on more complex likelihood functions.[9]

The source for virtually all the data on which the models were estimated was the Economic Council of Canada database.

A. Canadian Manufacturing Sector

The respective model parameters estimated using data for the Canadian manufacturing sector in aggregate are shown in Table 1. A comparison of the net and gross investment versions shows substantial differences in four parameters, with a_{0t} significantly larger (absolute) in the gross investment version (GIV), while the capital related coefficients a_{KK} and a_K differ in sign under GIV; moreover the latter coefficient becomes insignificant. These changes have implications for relationships among inputs.

More important is the fact that the coefficient a_{KK} under the net investment version (NIV) turns out to be negative (and significant). A positive a_{KK} is required to ensure a negative own price elasticity of capital, and for variable cost functions to be convex with respect to capital.

Another feature is that in the GIV the fit of the gross investment equation is markedly inferior to that for net investment in the net investment version. Seemingly this variation in fit reflects inclusion in the GIV of the (insignificant) coefficient a_{KK} applying to the ratio of gross investment to output—a relatively volatile quality—whereas in the NIV this term is constrained to be zero.

The size of the response parameter (M^*) is marginally higher in the NIV, suggesting that the overall rate of adjustment in the capital stock slows once account is taken of the way replacement investment may

Table 1. Estimated Parameters of Dynamic Model, Canadian Manufacturing, 1957–1982

Parameters	Gross Investment Version		Net Investment Version	
a_0	0.844	(3.11)	0.043	(0.74)
a_α	−0.012	(−3.60)	−0.005	(−5.85)
a_{EE}	−0.005	(−5.22)	−0.004	(−4.73)
a_{MM}	−0.033	(−4.74)	−0.03	(−4.65)
a_{KK}	17.08	(0.54)	−57.56	(−5.48)
$a_{\dot{K}\dot{K}}$	2619.80	(2.09)	−1901.80	(−10.12)
a_K	−1.99	(−0.64)	5.988	(5.46)
a_{Kt}	−0.030	(−1.57)	−0.023	(−2.00)
a_{EM}	0.005	(2.09)	0.003	(1.48)
a_K	−60.64	(−2.23)	—	—
$a_{\dot{K}t}$	0.575	(2.83)	—	—
$a_{\dot{K}K}$	116.75	(0.82)	—	—
a_E	0.010	(2.94)	0.013	(3.93)
a_{Et}	−0.00002	(−0.35)	−0.00007	(−1.13)
a_{EK}	0.088	(4.24)	0.069	(3.42)
$a_{E\dot{K}}$	−0.054	(−0.63)	—	—
a_{Mt}	−0.0003	(−1.34)	0.0004	(1.96)
a_{MK}	0.105	(1.51)	0.172	(2.37)
a_M	0.685	(62.74)	0.663	(62.89)
$a_{M\dot{K}}$	−1.357	(−3.82)	—	—
R^2 − L/Q	0.979		0.976	
− E/Q	0.837		0.835	
− M/Q	0.944		0.936	
− I_g/Q	0.055	Ǩ/Q	0.346	
M^*	0.139 (5.35)		0.176 (10.91)	
Ln Li	493.8		494.6	

Notes: t-values in parentheses.
L = labor; E = energy; M = intermediate materials; K = capital.

affect adjustment costs. The difference between the respective log of likelihood functions was not significant.

Differences do not arise between the models in terms of the nature of relationships between the inputs in the short run—all combinations of variable inputs were substitutes in both the NIV and GIV. However, substantive differences emerged over the long run. In particular, in the NIV complementarity over certain periods is shown in the case of (L) labor and energy (E), while in the GIV the two variables are always substitutes. Contrary results hold between capital (K) and labor: in the NIV they are complements, whereas under GIV they are substitutes. On the other hand, the NIV model finds substitutability between K and E, whereas the GIV finds complementarity.

Tables 2 and 3 show selected elasticities for the NIV and GIV models (calculated for actual rather than fitted values).[10] In the short run there is little variation, although the cross price elasticity of energy with respect

Table 2. Price Elasticities, Net Investment Version, Canadian Manufacturing (selected years)

Variable Input Price Elasticities

Year	E_{LL}	E_{LE}	E_{LM}	E_{EL}	E_{EE}	E_{EM}	E_{ML}	E_{ME}	E_{MM}
Short run									
1958	−0.212	0.190 E−01	0.193	0.216	−0.501	0.285	0.721 E−01	0.935 E−02	−0.814 E−01
1961	−0.201	0.101 E−01	0.191	0.124	−0.379	0.255	0.694 E−01	0.748 E−02	−0.769 E−01
1966	−0.181	0.675 E−02	0.174	0.895 E−01	−0.320	0.230	0.587 E−01	0.586 E−02	−0.646 E−01
1971	−0.125	0.497 E−02	0.121	0.720 E−01	−0.249	0.177	0.451 E−01	0.456 E−02	−0.496 E−01
1976	−0.131	0.145 E−01	0.116	0.159	−0.339	0.180	0.398 E−01	0.559 E−02	−0.454 E−01
1980	−0.191	0.281 E−01	0.163	0.237	−0.454	0.217	0.435 E−01	0.689 E−02	−0.504 E−01
1982	−0.195	0.489 E−01	0.146	0.330	−0.533	0.204	0.385 E−01	0.795 E−02	−0.465 E−01
Long run									
1958	−0.121	0.101 E−01	0.175	0.115	−0.491	0.305	0.655 E−01	0.100 E−0	−0.801 E−0
1961	−0.101	0.229 E−02	0.173	0.282 E−01	−0.372	0.273	0.628 E−01	0.799 E−02	−0.757 E−01
1966	−0.694 E−01	−0.049 E−02	0.155	−0.644 E−02	−0.314	0.246	0.525 E−01	0.627 E−01	−0.636 E−01
1971	−0.163 E−01	−0.108 E−02	0.105	−0.157 E−01	−0.244	0.189	0.394 E−01	0.488 E−02	−0.488 E−01
1976	−0.215 E−01	0.633 E−02	0.101	0.698 E−01	−0.333	0.193	0.346 E−01	0.598 E−02	−0.447 E−01
1980	−0.392 E−01	0.155 E−01	0.142	0.130	−0.445	0.232	0.379 E−01	0.736 E−02	−0.496 E−01
1982	−0.205 E−01	0.306 E−01	0.122	0.206	−0.523	0.218	0.321 E−01	0.849 E−02	−0.457 E−01

*Fixed Input Price Elasticities**

	E_{LK}	E_{EK}	E_{MK}	E_{KL}	E_{KE}	E_{KM}	E_{KK}
1958	−0.643 E−01	0.712 E−01	0.468 E−02	−0.218	0.212 E−01	0.424 E−01	0.154
1961	−0.746 E−01	0.712 E−01	0.494 E−02	−0.227	0.175 E−01	0.414 E−01	0.168
1966	−0.856 E−01	0.739 E−01	0.477 E−02	−0.264	0.172 E−01	0.436 E−01	0.203
1971	−0.880 E−01	0.707 E−01	0.456 E−02	−0.232	0.128 E−01	0.321 E−01	0.187
1976	−0.862 E−01	0.704 E−01	0.408 E−02	−0.215	0.159 E−01	0.298 E−01	0.169
1980	−0.118	0.827 E−01	0.442 E−02	−0.237	0.198 E−01	0.333 E−01	0.184
1982	−0.173	0.987 E−01	0.518 E−02	−0.226	0.191 E−01	0.257 E−01	0.181

Notes: L = labor; E = energy; M = materials; K = capital.
E_{LL} = own price elasticity of labor.
E_{LE} = cross price elasticity of labor quantity with respect to price of energy.
*Long-run elasticities only; short-run elasticities by definition are zero.

Table 3. Price Elasticities, Gross Investment Version, Canadian Manufacturing (selected years)

Variable Input Price Elasticities

Year	E_{LL}	E_{LE}	E_{LM}	E_{EL}	E_{EE}	E_{EM}	E_{ML}	E_{ME}	E_{MM}
Short run									
1958	-0.197	0.144 E-01	0.182	0.168	-0.606	0.439	0.689 E-01	0.143 E-01	-0.832 E-01
1961	-0.189	0.532 E-02	0.184	0.657 E-01	-0.452	0.386	0.672 E-01	0.114 E-01	-0.787 E-01
1966	-0.175	0.250 E-02	0.172	0.328 E-01	-0.388	0.355	0.572 E-01	0.897 E-02	-0.661 E-01
1971	-0.123	0.205 E-02	0.121	0.288 E-01	-0.299	0.270	0.438 E-01	0.698 E-02	-0.508 E-01
1976	-0.128	0.126 E-01	0.116	0.133	-0.410	0.276	0.379 E-01	0.855 E-02	-0.465 E-01
1980	-0.185	0.263 E-01	0.159	0.213	-0.544	0.331	0.411 E-01	0.105 E-01	-0.516 E-01
1982	-0.176	0.469 E-01	0.129	0.323	-0.628	0.305	0.353 E-01	0.121 E-01	-0.475 E-01
Long run									
1958	-0.196	0.143 E-01	0.183	0.240	-0.618	0.448	0.689 E-01	0.143 E-01	-0.832 E-01
1961	-0.192	0.563 E-02	0.183	0.135	-0.460	0.394	0.672 E-01	0.114 E-01	-0.787 E-01
1966	-0.182	0.323 E-02	0.171	0.105	-0.395	0.363	0.571 E-01	0.898 E-02	-0.661 E-01
1971	-0.136	0.311 E-02	0.119	0.975 E-01	-0.304	0.276	0.438 E-01	0.699 E-02	-0.508 E-01
1976	-0.146	0.145 E-01	0.114	0.204	-0.417	0.282	0.379 E-01	0.856 E-02	-0.465 E-01
1980	-0.219	0.304 E-01	0.156	0.296	-0.554	0.338	0.410 E-01	0.105 E-01	-0.516 E-01
1982	-0.236	0.539 E-01	0.125	0.423	-0.640	0.312	0.353 E-01	0.121 E-01	-0.475 E-01

Fixed Input Price Elasticities*

	E_{LK}	E_{EK}	E_{MK}	E_{KL}	E_{KE}	E_{KM}	E_{KK}
1958	-0.085 E-02	-0.700 E-01	0.005 E-02	0.127	-0.199 E-01	0.164 E-01	-0.123
1961	0.260 E-02	-0.688 E-01	0.006 E-02	0.134	-0.164 E-01	0.160 E-01	-0.134
1966	0.732 E-02	-0.728 E-01	0.006 E-02	0.162	-0.162 E-01	0.168 E-01	-0.162
1971	0.131 E-01	-0.689 E-01	0.005 E-02	0.149	-0.121 E-01	0.124 E-01	-0.149
1976	0.173 E-01	-0.690 E-01	0.005 E-02	0.138	-0.150 E-01	0.115 E-01	-0.135
1980	0.328 E-01	-0.804 E-01	0.005 E-02	0.153	-0.186 E-01	0.129 E-01	-0.147
1982	0.565 E-01	-0.944 E-01	0.006 E-02	0.152	-0.179 E-01	0.993 E-02	-0.144

Notes: L = labor; E = energy; M = materials; K = capital.
E_{LL} = own price elasticity of labor.
E_{LE} = cross price elasticity of labor quantity with respect to price of energy.
*Long-run elasticities only; short-run elasticities by definition are zero.

materials (E_{EM}) is consistently, although not appreciably, higher in the GIV.

In the long run, under the NIV the positive own price elasticity of capital perverts all the capital related elasticities including the short-long run relationships (absolute short run own price elasticities exceed the long run), contradicting the Le Chatelier Principle and showing that the boundary conditions of the model are transgressed.

Examination of the various short- and long-run output elasticities for the two models shows little systematic variation, with the exception that short-run labor-output elasticities are somewhat higher in the GIV. Estimates of short run multifactor productivity, in the form of cost diminution, are broadly consistent, although since the mid 1970s there has been a tendency for estimates of productivity in the GIV to exceed those from the NIV.

B. Canadian Iron and Steel Industry

The respective model parameters are shown in Table 4. Substantive variations in common parameters were confined to capital related coefficients ($a_{\bar{K}\bar{K}}$, a_{KK}, a_K, a_{Kt}, and a_{MK}). Generally the equation fit was superior in the GIV; the difference between the values of the respective (log) likelihood functions was substantial. The speed of response coefficient (M^*) was estimated to be markedly higher in the NIV than in the GIV. This suggests that some of the variations attributable to replacement investment in the gross investment formulation are being captured by this parameter; in particular lower imputed adjustment costs in the NIV would tend to accelerate responses.

The relationships between inputs in the models show little variation. In the short run, some complementarity was shown between labor and materials in the GIV, but not in the NIV. In the long run, the character of virtually all relationships did not vary between specifications, with the exceptions of labor-capital, which were substitutes under NIV, but sometimes complements in GIV.

Tables 5 and 6 show selected elasticities for the NIV and GIV models. For variable inputs, the main difference was the higher (absolute) "own" price elasticities of labor in both the short and long run under NIV; also in the short run the energy-labor cross price elasticity is higher (absolute) in the NIV. The labor-capital elasticities are noticeably positive in the NIV; their magnitude in the GIV is trivial.

Table 4. Estimated Parameters of Dynamic Model, Iron, and Steel, 1957–1982

Parameters	Gross Investment Version		Net Investment Version	
a_0	0.787	(8.63)	0.744	(9.00)
a_α	−0.009	(−3.07)	−0.007	(−2.72)
a_{EE}	−0.045	(−7.02)	−0.047	(−7.00)
a_{MM}	−0.065	(−4.17)	−0.085	(−5.72)
a_{KK}	7.91	(1.33)	10.98	(3.62)
$a_{\dot{K}\dot{K}}$	194.76	(3.83)	67.91	(3.65)
a_K	−1.737	(−1.94)	−2.743	(−3.95)
a_{Kt}	0.010	(0.78)	−0.007	(−0.96)
a_{EM}	0.053	(6.94)	0.054	(6.84)
$a_{\dot{K}}$	−10.034	(−2.07)	—	—
$a_{\dot{K}t}$	−0.098	(−1.19)	—	—
$a_{K\dot{K}}$	8.622	(0.41)	—	—
a_E	−0.039	(−5.21)	−0.039	(−5.15)
a_{Et}	0.002	(7.22)	0.002	(7.04)
a_{EK}	0.189	(7.42)	0.231	(9.30)
$a_{E\dot{K}}$	0.295	(4.32)	—	—
a_{Mt}	−0.0001	(−0.14)	−0.00009	(−0.13)
a_{MK}	−0.452	(−5.61)	−0.774	(−13.50)
a_M	0.713	(34.19)	0.734	(39.65)
$a_{M\dot{K}}$	−2.552	(−10.23)	—	—
R^2 − L/Q	0.899		0.675	
− E/Q	0.477		0.418	
− M/Q	0.840		0.402	
− I_g/Q	0.152		K̇/Q 0.263	
M^*	0.23 (7.73)		0.390 (8.63)	
Ln Li	370.6		348.9	

Notes: t-values in parentheses.
L = labor; E = energy; M = intermediate materials; K = capital.

The variations between short and long run output elasticities for the two models are not appreciable; the same comment applies to estimates of multifactor productivity.

C. Canadian Textiles

Parameter estimates are shown in Table 7. Similarly to iron and steel, the main differences appear in the capital related coefficients (a_{KK}, $a_{\dot{K}\dot{K}}$, a_K, a_{Kt}, and a_{MK}) which generally have a higher absolute magnitude in the GIV; moreover, the parameter a_{MK} switches in sign under GIV, while retaining significance. The equation fits are much the same with the exception of the materials equation, which shows a marked improvement under GIV. The difference between the (log) likelihood functions was substantial, with the GIV model having a considerably larger sample value. The response coefficient (M^*) is noticeably higher

Table 5. Price Elasticities, Net Investment Version, Iron and Steel (selected years)

Variable Input Price Elasticities

Year	E_{LL}	E_{LE}	E_{LM}	E_{EL}	E_{EE}	E_{EM}	E_{ML}	E_{ME}	E_{MM}
Short run									
1958	−0.168	−0.589 E−01	0.227	−0.400	−2.075	2.475	0.118	0.190	−0.308
1961	−0.134	−0.527 E−01	0.186	−0.431	−1.777	2.208	0.100	0.145	−0.245
1966	−0.108	−0.402 E−01	0.149	−0.378	−1.766	2.144	0.793 E−01	0.122	−0.201
1971	−0.706 E−01	−0.212 E−01	0.918 E−01	−0.210	−1.410	1.620	0.545 E−01	0.968 E−01	−0.151
1976	−0.933 E−01	0.472 E−01	0.462 E−01	0.263	−1.829	1.565	0.221 E−01	0.134	−0.156
1980	−0.108	0.205 E−01	0.871 E−01	0.980 E−01	−1.603	1.505	0.358 E−01	0.129	−0.165
1982	−0.112	0.940 E−01	0.183 E−01	0.315	−1.411	1.095	0.826 E−02	0.147	−0.156
Long run									
1958	−0.168	−0.556 E−01	0.215	−0.378	−2.289	3.220	0.112	0.247	−0.506
1961	−0.141	−0.399 E−01	0.140	−0.327	−1.961	2.873	0.751 E−01	0.188	−0.403
1966	−0.148	−0.126 E−01	0.510 E−01	−0.119	−1.949	2.789	0.273 E−01	0.158	−0.331
1971	−0.128	0.788 E−02	−0.538 E−02	0.783 E−01	−1.556	2.108	−0.319 E−02	0.126	−0.249
1976	−0.143	0.882 E−01	−0.562 E−01	0.492	−2.017	2.036	−0.269 E−01	0.175	−0.257
1980	−0.224	0.839 E−01	−0.865 E−01	0.402	−1.769	1.958	−0.355 E−01	0.168	−0.271
1982	−0.341	0.194	−0.207	0.650	−1.557	1.425	−0.934 E−01	0.192	−0.256

Fixed Input Price Elasticities*

	E_{LK}	E_{EK}	E_{MK}	E_{KL}	E_{KE}	E_{KM}	E_{KK}
1958	0.832 E−02	−0.552	0.147	0.139 E−01	−0.135	0.470	−0.349
1961	0.409 E−01	−0.586	0.139	0.711 E−01	−0.125	0.451	−0.398
1966	0.109	−0.722	0.145	0.189	−0.133	0.470	−0.526
1971	0.126	−0.630	0.126	0.226	−0.114	0.382	−0.493
1976	0.111	−0.511	0.109	0.153	−0.126	0.314	−0.341
1980	0.227	−0.591	0.139	0.266	−0.145	0.397	−0.518
1982	0.354	−0.518	0.158	0.260	−0.114	0.257	−0.404

Notes: L = labor; E = energy; M = materials; K = capital.
E_{LL} = own price elasticity of labor.
E_{LE} = cross price elasticity of labor quantity with respect to price of energy.
*Long-run elasticities only; short-run elasticities by definition are zero.

Table 6. Price Elasticities, Gross Investment Version, Iron and Steel (selected years)

Variable Input Price Elasticities

Year	E_{LL}	E_{LE}	E_{LM}	E_{EL}	E_{EE}	E_{EM}	E_{ML}	E_{ME}	E_{MM}
Short run									
1958	−0.275 E-01	−0.578 E-01	0.853 E-01	−0.470	−2.178	2.648	0.454 E-01	0.173	−0.219
1961	−0.259 E-01	−0.523 E-01	0.782 E-01	−0.476	−1.787	2.263	0.434 E-01	0.138	−0.181
1966	−0.212 E-01	−0.441 E-01	0.653 E-01	−0.398	−1.680	2.079	0.341 E-01	0.120	−0.154
1971	−0.191 E-01	−0.251 E-01	0.371 E-01	−0.226	−1.325	1.551	0.209 E-01	0.970 E-01	−0.118
1976	−0.129 E-01	0.364 E-01	−0.235 E-01	0.233	−1.829	1.596	−0.122 E-01	0.129	−0.117
1980	−0.127 E-01	0.150 E-01	−0.230 E-02	0.660 E-01	−1.528	1.462	−0.088 E-02	0.127	−0.126
1982	−0.182 E-01	0.563 E-01	−0.382 E-01	0.324	−1.555	1.230	−0.235 E-01	0.132	−0.108
Long run									
1958	−0.303 E-01	−0.748 E-01	0.149	−0.501	−2.369	3.359	0.522 E-01	0.215	−0.372
1961	−0.235 E-01	−0.606 E-01	0.110	−0.431	−1.944	2.871	0.339 E-01	0.171	−0.309
1966	−0.238 E-01	−0.418 E-01	0.568 E-01	−0.231	−1.828	2.637	0.164 E-02	0.149	−0.262
1971	−0.196 E-01	−0.206 E-01	0.209 E-01	−0.278 E-01	−1.441	1.968	−0.185 E-01	0.120	−0.200
1976	−0.111 E-01	0.346 E-01	−0.188 E-01	0.396	−1.990	2.025	−0.433 E-01	0.160	−0.199
1980	−0.329 E-01	0.275 E-01	−0.391 E-01	0.282	−1.662	1.855	−0.495 E-01	0.157	−0.214
1982	−0.635 E-01	0.779 E-01	−0.903 E-01	0.612	−1.691	1.561	−0.893 E-01	0.163	−0.184

Fixed Input Price Elasticities*

	E_{LK}	E_{EK}	E_{MK}	E_{KL}	E_{KE}	E_{KM}	E_{KK}
1958	−0.435 E-01	−0.489	0.105	−0.194 E-01	−0.119	0.442	−0.304
1961	−0.267 E-01	−0.496	0.104	0.317 E-01	−0.109	0.424	−0.346
1966	0.884 E-02	−0.579	0.112	0.133	−0.117	0.442	−0.458
1971	0.193 E-01	−0.499	0.988 E-01	0.171	−0.100	0.359	−0.430
1976	−0.472 E-02	−0.431	0.822 E-01	0.112	−0.111	0.295	−0.297
1980	0.445 E-01	−0.475	0.107	0.205	−0.127	0.373	−0.451
1982	0.760 E-01	−0.481	0.110	0.210	−0.997 E-01	0.241	−0.352

Notes: L = labor; E = energy; M = materials; K = capital.
E_{LL} = own price elasticity of labor.
E_{LE} = cross price elasticity of labor quantity with respect to price of energy.
*Long-run elasticities only; short-run elasticities by definition are zero.

181

Table 7. Estimated Parameters of Dynamic Model, Textiles, 1957–1982

Parameters	Gross Investment Version		Net Investment Version	
a_0	0.835	(10.23)	0.620	(9.64)
a_{0t}	−0.023	(−7.09)	−0.015	(−6.27)
a_{EE}	−0.005	(−3.69)	−0.006	(−4.07)
a_{MM}	−0.111	(−17.55)	−0.102	(−14.57)
a_{KK}	29.38	(4.87)	14.80	(3.10)
$a_{\dot{K}\dot{K}}$	664.97	(7.40)	206.97	(2.84)
a_K	−3.62	(−3.96)	−2.52	(−3.00)
a_{Kt}	0.084	(3.41)	−0.024	(1.18)
a_{EM}	0.007	(3.40)	0.008	(3.83)
$a_{\dot{K}}$	−10.46	(−1.72)	—	—
$a_{\dot{K}t}$	0.029	(0.18)	—	—
$a_{\dot{K}K}$	−86.65	(−2.65)	—	—
a_E	0.001	(0.27)	0.001	(0.30)
a_{Et}	0.0003	(2.36)	0.0004	(2.48)
a_{EK}	0.066	(6.90)	0.063	(6.59)
$a_{E\dot{K}}$	−0.011	(−0.44)	—	—
a_{Mt}	−0.004	(−6.87)	−0.007	(−10.49)
a_{MK}	0.219	(3.38)	−0.662	(−11.58)
a_M	0.765	(45.48)	0.903	(52.87)
$a_{M\dot{K}}$	−0.053	(−0.14)	—	—
R^2 – L/Q	0.994		0.973	
– E/Q	0.978		0.977	
– M/Q	0.859		0.667	
– I_g/Q	0.318	$\dot{K}$/Q	0.341	
M^*	0.129 (5.84)		0.257 (6.37)	
Ln Li	445.5		409.6	

Notes: t-values in parentheses.
L = labor; E = energy; M = intermediate materials; K = capital.

under NIV than under GIV, again suggesting it may be picking up changes associated with replacement investment that are separately identified in the GIV, and that lower adjustment costs calculated under the NIV accelerate responses in quasi-fixed inputs.

The nature of the relationships among inputs is the same between the two models in the short run. For the long run, although the relationships between variable inputs are similar, in the case of capital there is one significant change—capital and materials are substitutes in the NIV and complements in the GIV—and one modest change: some capital-labor complementarity is shown in the NIV, while the two inputs are always substitutes under GIV.

The magnitudes of the elasticities under the two models are shown in Tables 8 and 9. The short run elasticities are much the same, but differences emerge for the long run elasticities. For variable inputs, the labor-materials long run cross elasticity is smaller under NIV in the second half of the estimation period, whereas the materials-labor elas-

Table 8. Price Elasticities, Net Investment Version, Textiles (selected years)

Variable Input Price Elasticities

Year	E_{LL}	E_{LE}	E_{LM}	E_{EL}	E_{EE}	E_{EM}	E_{ML}	E_{ME}	E_{MM}
Short run									
1958	−0.705	−0.167 E-01	0.722	−0.242	−0.594	0.836	0.395	0.315 E-01	−0.427
1961	−0.671	−0.179 E-01	0.689	−0.302	−0.510	0.813	0.330	0.230 E-01	−0.353
1966	−0.534	−0.143 E-01	0.548	−0.264	−0.438	0.702	0.247	0.171 E-01	−0.264
1971	−0.292	−0.664 E-02	0.298	−0.142	−0.380	0.522	0.146	0.120 E-01	−0.158
1976	−0.268	0.105 E-01	0.267	0.147 E-01	−0.461	0.446	0.121	0.146 E-01	−0.136
1980	−0.327	0.110 E-01	0.316	0.115	−0.602	0.487	0.122	0.180 E-01	−0.140
1982	−0.353	0.397 E-01	0.314	0.296	−0.771	0.475	0.112	0.228 E-01	−0.135
Long run									
1958	−0.720	−0.220 E-01	0.780	−0.319	−0.623	1.142	0.427	0.430 E-01	−0.551
1961	−0.673	−0.197 E-01	0.711	−0.333	−0.534	1.110	0.340	0.314 E-01	−0.455
1966	−0.542	−0.114 E-01	0.512	−0.210	−0.459	0.959	0.231	0.234 E-01	−0.341
1971	−0.326	−0.122 E-02	0.241	−0.261 E-01	−0.399	0.714	0.118	0.163 E-01	−0.204
1976	−0.317	0.981 E-02	0.202	0.136	−0.483	0.610	0.918 E-01	0.199 E-01	−0.176
1980	−0.451	0.295 E-01	0.201	0.308	−0.631	0.665	0.779 E-01	0.246 E-01	−0.181
1982	−0.590	0.739 E-01	0.152	0.551	−0.808	0.649	0.544 E-01	0.311 E-01	−0.174

Fixed Input Price Elasticities*

	E_{LK}	E_{EK}	E_{MK}	E_{KL}	E_{KE}	E_{KM}	E_{KK}
1958	−0.377 E-01	−0.200	0.814 E-01	−0.112	−0.412 E-01	0.444	−0.291
1961	−0.182 E-01	−0.242	0.837 E-01	−0.547 E-01	−0.431 E-01	0.526	−0.428
1966	0.409 E-01	−0.290	0.871 E-01	0.111	−0.428 E-01	0.526	−0.594
1971	0.861 E-01	−0.289	0.696 E-01	0.257	−0.404 E-01	0.425	−0.642
1976	0.105	−0.263	0.638 E-01	0.257	−0.464 E-01	0.345	−0.555
1980	0.221	−0.343	0.787 E-01	0.456	−0.678 E-01	0.420	−0.806
1982	0.365	−0.392	0.888 E-01	0.489	−0.706 E-01	0.333	−0.752

Notes: L = labor; E = energy; M = materials; K = capital.
E_{LL} = own price elasticity of labor.
E_{LE} = cross price elasticity of labor quantity with respect to price of energy.
*Long-run elasticities only; short-run elasticities by definition are zero.

Table 9. Price Elasticities, Gross Net Investment Version, Textiles (selected years)

Table 9. Price Elasticities, Gross Net Investment Version, Textiles (selected years)

Variable Input Price Elasticities

Year	E_{LL}	E_{LE}	E_{LM}	E_{EL}	E_{EE}	E_{EM}	E_{ML}	E_{ME}	E_{MM}
Short run									
1958	−0.715	−0.147 E−01	0.730	−0.231	−0.569	0.799	0.404	0.282 E−01	−0.432
1961	−0.745	−0.174 E−01	0.762	−0.291	−0.493	0.784	0.360	0.221 E−01	−0.382
1966	−0.593	−0.139 E−01	0.606	−0.255	−0.425	0.680	0.275	0.168 E−01	−0.292
1971	−0.356	−0.707 E−02	0.364	−0.137	−0.370	0.507	0.164	0.118 E−01	−0.176
1976	−0.303	0.108 E−02	0.302	0.047 E−01	−0.446	0.431	0.132	0.139 E−01	−0.146
1980	−0.343	0.102 E−01	0.333	0.112	−0.582	0.470	0.134	0.173 E−01	−0.152
1982	−0.275	0.272 E−01	0.247	0.284	−0.737	0.453	0.120	0.211 E−01	−0.141
Long run									
1958	−0.783	−0.868 E−02	0.750	0.450 E−01	−0.593	0.717	0.437	0.253 E−01	−0.442
1961	−0.845	−0.105 E−01	0.788	0.167 E−01	−0.514	0.704	0.393	0.198 E−01	−0.396
1966	−0.730	−0.642 E−02	0.635	0.809 E−01	−0.443	0.610	0.308	0.150 E−01	−0.299
1971	−0.477	−0.251 E−02	0.379	0.178	−0.386	0.455	0.188	0.105 E−01	−0.180
1976	−0.379	0.615 E−02	0.314	0.302	−0.465	0.387	0.154	0.125 E−01	−0.150
1980	−0.541	0.237 E−01	0.359	0.477	−0.607	0.421	0.161	0.155 E−01	−0.155
1982	−0.516	0.459 E−01	0.275	0.692	−0.768	0.406	0.148	0.189 E−01	−0.144

Fixed Input Price Elasticities*

Year	E_{LK}	E_{EK}	E_{MK}	E_{KL}	E_{KE}	E_{KM}	E_{KK}
1958	0.418 E−01	−0.169	−0.202 E−01	0.392	−0.347 E−01	−0.117	−0.240
1961	0.669 E−01	−0.206	−0.223 E−01	0.529	−0.363 E−01	−0.139	−0.354
1966	0.101	−0.248	−0.236 E−01	0.666	−0.361 E−01	−0.139	−0.491
1971	0.710 E−01	−0.247	−0.190 E−01	0.676	−0.341 E−01	−0.112	−0.530
1976	0.592 E−01	−0.224	−0.168 E−01	0.588	−0.392 E−01	−0.909 E−01	−0.458
1980	0.158	−0.292	−0.209 E−01	0.836	−0.572 E−01	−0.111	−0.668
1982	0.195	−0.330	−0.228 E−01	0.769	−0.595 E−01	−0.880 E−01	−0.621

Notes: L = labor; E = energy; M = materials; K = capital.
E_{LL} = own price elasticity of labor.
E_{LE} = cross price elasticity of labor quantity with respect to price of energy.
*Long-run elasticities only; short-run elasticities by definition are zero.

ticity is somewhat larger under GIV during this interval. The longer run energy-materials elasticity is larger under NIV. Several capital elasticities sow significant changes. In particular, labor-capital is larger in later years under NIV, whereas capital-labor is smaller under GIV. Capital-materials are noticeable substitutes under NIV, whereas they are complements under GIV. Although these sign changes also hold for materials-capital under both models, the elasticities are trivial in magnitude.

The estimates of multi-factor productivity are higher under the NIV than under GIV, but not appreciably so. There are some differences in short- and long-run labor-output elasticities; short run materials-output elasticities are modestly higher under NIV.

IV. CONCLUDING REMARKS

We confess a theoretical preference for expressing adjustment costs of the "internal" variety as a function of gross rather than net investment. The latter formulation has the undesirable (and unsupported) inherent assumption of treating replacement investment as frictionless. However, the gross investment specification is markedly more complex and can result in more difficult and even intractable estimation problems. Thus, a decision about the merits of the alternative specifications in an applied context hinges critically on empirical differences, including the significance of adjustment costs associated with replacement investment.[11]

The broad conclusions which emerge from the three sets of results analyzed under the net and gross investment versions can be summarized in terms of the degree of fit, speed of response, bias in technical change, the nature of relationships between inputs, and input elasticities.

The model for aggregate Canadian manufacturing showed some differences in equation fit, with an inferior performance for the investment equation in the gross investment specification; but for the two individual industry models the degree of fit would seem to be little affected by the net or gross investment specification, with even some improvement under the latter.

The rapidity with which quasi-fixed inputs move to equilibrium is governed by the response coefficient, M^*. In all cases, the coefficient was higher under the net investment formulation. This suggests the latter model is picking up in terms of speed of response changes which really emanate from the impact replacement investment has on the way factor inputs adjust over time. Moreover, the result confirms intuition that higher adjustment costs will curtail the speed of response; adjustment costs are lower in the net investment specification.

Variations in the bias of technical change according to model specification were not prominent, and for those which were recorded no clear pattern emerged.

In terms of changes in relationships between factors, a distinction can be drawn between those affecting variable inputs and those affecting quasi-fixed inputs. For variable inputs, virtually all relationships were consistent, especially in the short run. In the long run some differences in the magnitude of elasticities did emerge, mainly involving labor.

As expected a priori, the strongest differences between the models were recorded by the fixed input related coefficients, which in turn affected both the sign and magnitude of elasticities associated with fixed inputs. Differences in relationships were recorded, with some inputs recorded as substitutes under one specification appearing as complements under the alternative specification. Such differences in sign were mitigated sometimes by the small magnitude of the elasticities, but in other cases the spread between the elasticities was substantial. In the case of aggregate manufacturing, the net investment specification transgressed regularity conditions for the quasi-fixed input.

Estimates of output elasticities were little affected by the alternative specifications. However, there was some tendency for estimates of multifactor productivity to be greater for the gross investment model.

Overall, we conclude that significant differences emerge according to whether adjustment costs reflect net or gross investment-that such variations in specification do matter. In the main, any attempt to estimate a simpler third generation model by confining adjustment costs to be dependent on net investment carries with it the likelihood that the speed with which quasi-fixed factors move toward equilibrium levels will be larger in the case of individual industry models, that the estimated long run responses will be smaller, that the characteristic relationships between quasi-fixed inputs and certain variable inputs could be erroneous, and that regularity conditions might not be satisfied. There is also the possibility that productivity growth may be smaller.

Thus, our theoretical inclination for the gross investment specification becomes more emphatic in the light of empirical differences between the net and gross investment models. However, estimation of the more complex gross investment third generation models is more convoluted, and the simpler net investment version need not distort relationships, especially in the short run.

ACKNOWLEDGMENTS

The authors gratefully acknowledge helpful comments by John Moroney of Texas A&M University. The usual disclaimers apply.

NOTES

1. For example, see Berndt, Fuss, and Waverman (1979), Berndt and Morrison (1981), and Pindyck and Rotemberg (1983).
2. Note this relationship holds essentially for absolute values of $\dot{x}(t)$; it does not entail increasing output if $\dot{x}(t)$ became negative, implying disinvestment.
3. We caution that the assumption of strict convexity of the internal cost of adjustment function could be easily transgressed: the issue is essentially empirical.
4. In terms of empirical work, this assumption is important; see later.
5. White collar (nonproduction) can also be treated as quasi-fixed (see Berndt, Morrison, and Watkins, 1981) but lack of data precluded breaking down labor inputs in this way.
6. This form has three desirable properties: it constrains the second "cross" partial derivatives, G_{xx}^*, to be symmetric (a theoretical requirement for the model specification); the Hessians of second order partial derivatives constitute a matrix of constants, thus facilitating the linking of short and long-run responses; and the characterization of the optimal path for the quasi-fixed factor is globally as well as locally valid (Treadway, 1971).
7. The form of the expression (16) attempts to impose overall long run constant returns to scale (LRCRTS). However, it is only an approximation; see Watkins (1987).
8. Note that the K^* equilibrium capital stock is associated with a fixed exogenous rate of output Q; thus in equilibrium the capital : output ratio, K^*/Q, is constant.
9. However, the standard errors of the estimates are biased; see Prucha and Nadiri (1986) and Prucha (1987).
10. Many of the elasticity expressions—and especially those for the long run — are too complex to allow easy derivation of their standard errors. Because the variances of the elasticities have not been evaluated, we caution that the recorded differences discussed below need not be statistically significant.
11. Tests of the significance of the replacement investment component of gross investment in terms of adjustment costs could be undertaken by respecifying Equation 16 to have different coefficients attached to ΔK and dK_{-1}. This development was not pursued here. Moreover, empirical estimates of the magnitude of adjustment costs under the net and gross investment formulations have not been undertaken in this paper.

REFERENCES

Berndt, E.R., M.A. Fuss, and L. Waverman. 1979. "A Dynamic Model of Cost Adjustment and Interrelated Factor Demands." Working Paper No. 7925. Toronto: Institute for Policy Analysis, University of Toronto.

Berndt, E.R., and C. Morrison. 1981. "Short Run Labor Productivity in a Dynamic Model." *Journal of Econometrics*, 16.

Berndt, E.R., C. Morrison, and G.C. Watkins. 1981. "Dynamic Models of Energy Demand: An Assessment and Comparison." In *Modeling and Measuring Natural Resources Substitution*, edited by E.R. Berndt and B. Field. Cambridge, MA: MIT Press.

Jorgensen, D. 1972. "Investment Behaviour and the Production Function." *Bell Journal of Economics and Management Science* (Spring).

Lau, L. 1976. "A Characterization of the Normalized Restricted Profit Function." *Journal of Economic Theory*, 12.

Lucas, R.E. 1967. "Optimal Investment Policy and Flexible Accelerator. *International Economic Review*, 8, 1.

Morrison, C. 1982. "Three Essays on the Dynamic Analysis of Demand for Factors of Production." Unpublished Ph.D. Dissertation, University of British Columbia, Department of Economics.

Pindyck, R.S., and J.J. Rotemberg. 1983. "Dynamic Factor Demands and the Effects of Energy Price Shocks." *The American Economic Review*, 73, 5.

Prucha, I.R. 1987. "The Variance-Covariance Matrix of the Maximum Likelihood Estimation in Triangular Structural Systems; Consistent Estimation." *Econometrica*, 55, 4: 977–978.

Prucha, I.R., and M.I. Nadiri, 1986. "A Comparison of Alternative Methods for the Estimation of Dynamic Factor Demand Models Under Non-Static Expectations." *Journal of Econometrics: The Econometrics of Temporary Equilibrium*, 33 (Supplement).

Treadway, A.B. 1971. "The Rational Multivariate Flexible Accelerator." *Econometrica*, 39, 5.

Treadway, A.B. 1974. "Globally Optical Flexible Accelerator." *Journal of Economic Theory*, 7.

Watkins, G.C. 1984. "Dynamic Models of Industrial Energy Demand." Mimeo.

Watkins, G.C. 1987. "Note on Returns to Scale When Adjustment Costs are a Function of Gross Investment." *University of Calgary Discussion Paper Series No. 105* (March).

ENERGY, CAPITAL, AND GROWTH

John R. Moroney

ABSTRACT

This paper presents estimates of a two-equation time-series model of energy demand, capital, and real income growth for the United States, Canada, Sweden, West Germany, and France. Per capita demand for energy depends on real per capita income and the real price of energy. Own price elasticities are found to be significantly less than one, while income elasticities are approximately one. Real per capita income depends on physical capital and energy consumption per head. The pervasive reduction in per capita growth following the surge in real energy prices beginning in 1973 is attributable chiefly to reduced energy consumption.

I. INTRODUCTION

Two views now dominate scientific thought concerning the determinants of long-run per capita incomes. The first, stressed chiefly by economists, is that real incomes depend primarily on physical capital and its embodied technology per person.[1] Alternatively, physical scientists ascribe primary importance to the per capita energy expended to locate, extract, and process nonrenewable resources (Cook 1971; Goeller and Weinberg 1976; Bockris 1980; Cleveland, Costanza, Hall, and Kaufman 1984).[2] Using time-series samples for five industrial countries—France, the Federal Republic of Germany, Sweden, Canada, and the United States—this paper demonstrates that both capital and energy are vitally important. Economies driven by capital accumulation alone, unaccompanied by rising energy consumption, display sharply lower real income growth than those marked by increasing capital and energy.

Physical capital, energy, and nonfuel minerals are of course closely linked in industrial societies. The major nonfuel minerals, iron and aluminum oxides, calcium carbonates, and silicon oxides, are the building blocks of physical capital. Energy is essential both to transform these minerals into new capital and to activate the capital already in place. From an economic viewpoint, however, capital and energy are distinct resources whose influences on real income can be analyzed separately: Societies can increase capital per person without raising energy consumption, and vice versa. Indeed, during the period of abruptly increasing energy prices (1973–1981), France, Germany, Sweden, and the United States curtailed per capita energy consumption yet continued to accumulate real capital per person.

Current time-series evidence suggests that physical capital and energy are broadly complementary, but certainly need not be utilized in fixed proportions.[3] By contrast, international studies based on pooled cross-section, time-series samples indicate considerable substitutability between capital and energy.[4] Yet a question of the first importance has not been addressed in these or other studies: What are the separate contributions of capital and energy per person in promoting per capita income growth? If energy plays a secondary and capital a primary role, public policy should chiefly center on the enhancement of real investment. But if energy were consequential in its own right, policies should rightly focus on measures to augment future energy supplies.

I propose a simple, two-equation model to analyze two related issues.

1. An energy demand equation is employed to analyze the effects of energy prices and per capita incomes on per capita energy consumption.
2. An income growth equation then provides estimates of the

separate contributions of physical capital and energy consumption per capita to the growth of real per capita incomes.[5]

The equations are estimated simultaneously by three-stage least-squares.

The model is applied to annual time series for five nations in the Organization for Economic Cooperation and Development: Canada, France, and the United States for the period 1955–1980; Sweden for the years 1962–1980; and West Germany for the years 1959–1980. The shorter periods of analysis for Sweden and West Germany were dictated by the availability of real capital stocks.

II. AN OVERVIEW: PER CAPITA INCOME, CAPITAL, AND ENERGY

It is useful at the outset to summarize the salient trends in per capita income, capital, and energy consumption. First some definitions. Per capita income in each country is measured as real gross domestic product per capita (y_t) expressed in U.S. $1975, adjusted for changes in purchasing power parity. These estimates are comparable across countries and over time. They are the major objective result of the World Bank International Comparison Project, Phase III, and are obtained from Summers and Heston (1984).[6]

Energy consumption per capita (E_t) is obtained from various issues of the United Nations *Yearbook of World Energy Statistics*. Energy is measured comprehensively from all sources: coal and lignite, petroleum, natural gas and natural gas liquids, hydroelectricity, and nuclear-based electricity. All such energy is converted to kilograms of standard coal equivalents (7,000 calories per gram), hence is fully comparable across countries and over time.[7] This comprehensive measure explicitly accounts for differences across countries and changes over time within countries in the composition of energy inputs.

Real capital per person (k_t) is the aggregate gross value of structures and equipment in all sectors (agriculture, industry, services, dwellings, and government), expressed in constant prices, and divided by population. The constant-price gross capital stocks are developed using a standard perpetual inventory method described in Ward (1976). The estimates are obtained from the OECD *Flows and Stocks of Fixed Capital* (1983).

Growth rates of per capita income, capital, and energy consumption

Table 1. Average Annual Rates of Growth of Per Capita Income, Capital, and Energy in Five OECD Countries (In Percent)

Country	Time Period	Per Capita Income	Capital Stock Per Capita	Energy Consumption Per Capita
France	1955–1973	4.7	3.5	4.0
	1973–1980	1.8	4.3	−0.2
Germany	1959–1973	4.1	4.6	4.0
	1973–1980	2.1	3.7	−0.3
Sweden	1962–1973	3.1	3.5	4.1
	1973–1980	0.9	3.0	−1.3
Canada	1955–1973	3.1	3.1	3.3
	1973–1980	1.7	3.8	1.1
United States	1955–1973	2.2	2.1	2.2
	1973–1980	0.8	2.5	−1.1

Sources: Per capita income estimates are from Summers and Heston (1984).
Capital stocks are from Organization for Economic Cooperation and Development (1983) and population estimates from Summers and Heston (1984).
Energy consumption estimates are from United Nations, *Yearbook of World Energy Statistics* (various issues).

are shown in Table 1. Note that all countries except the United States displayed rapid growth in real per capita income until 1973, when energy prices began to increase abruptly. During the years 1973–1980, however, per capita income growth diminished sharply to one-half or less of its earlier rate, except in Canada where the growth rate declined from 3.1 percent to 1.7 percent.

Remarkably, until 1973 each country exhibited very nearly balanced growth in capital and energy consumption per capita. And both rates closely paralleled the country's own growth in per capita income. Then things changed dramatically. Between 1973 and 1980, France, Germany, Sweden, and the United States actually reduced per capita energy consumption; Canada continued to expand, but at a much diminished rate of 1.1 percent annually.

Following the energy price shocks that began in 1973, capital per head continued to increase in all countries. Indeed, the growth rates accelerated in France, Canada, and the United States. Yet capital deepening, by itself, could not stem steep declines in the growth of real income per head.

Stationary or declining energy consumption per head is a major cause of the pervasive post-1973 decrease in per capita income growth. Capital accumulation is intrinsically important; but so is energy consumption. To ignore either as a source of real income growth is to err fundamentally.

III. SPECIFICATION

Aggregate desired per capita demand for energy depends on historical real energy prices and real income per capita. Additional factors such as differences in average temperature and population densities should probably be accounted for in international cross-section studies, but are unimportant in time-series analyses of specific countries. Accordingly, we specify the desired demand for energy in period t as

$$\ln E_t^* = g(P_t, y_t) = \alpha_0 + \alpha_1 \ln P_t + \alpha_2 \ln y_t \qquad (1)$$

where E_t^* is desired per capita energy demand, P_t is the real price of energy, and y_t is real per capita income. Note that Equation (1) contains no technical or behavioral disturbance term since it specifies only a desired relationship. Energy is used in modern societies to produce new capital and to activate existing capital goods. Thus, equilibrium energy consumption is unlikely to adjust fully to price and income changes within one year, since capital stocks themselves adjust only partially within a year. Yet the desired flow of energy is considerably more adaptable than long-run equilibrium stocks of capital. Thus, we specify an adjustment process[8]

$$E_t/E_{t-1} = (E_t^*/E_{t-1})^\delta, \; 0 \le \delta \le 1 \qquad (2)$$

where E_t is actual energy consumption and δ is the coefficient of adjustment. Taking the logarithm of Equation (2), substituting Equation (1), and appending a disturbance, one obtains an energy demand equation

$$\ln E_t = \beta_0 + \beta_1 \ln P_t + \beta_2 \ln y_t + \beta_3 \ln E_{t-1} + \varepsilon_t \qquad (3)$$

where $\beta_0 = \delta\alpha_0$, $\beta_1 = \delta\alpha_1$, $\beta_2 = \delta\alpha_2$, and $\beta_3 = (1 - \delta)$. The short-run (one-year) and long-run price elasticities of demand are β_1 and $\beta_1^* = \beta_1/(1 - \beta_3)$; the corresponding income elasticities are β_2 and $\beta_2^* = \beta_2/(1 - \beta_3)$. It should be stressed that if $\ln P_t$ and $\ln y_t$ are predetermined instrumental variables and ε_t is not autocorrelated, consistent estimates of these parameters are obtained by applying ordinary least squares to Equation (3) (Kmenta, 1986, pp. 535–536). If, however, the original disturbance ε_t follows a first-order autoregressive process $\varepsilon_t = \rho\varepsilon_{t-1} + \mu_t$ where $\mu_t \sim \text{NID}(0, \sigma_\mu^2)$ and $E(\mu_t\varepsilon_{t-1}) = 0$, maximum likelihood estimates of (3) can be obtained by an iterative search for the maximum likelihood value of $|\rho| < 1$ (Zellner and Geisel, 1970).

A realistic examination of the roles of energy and capital in real income growth calls for a functional form permitting these factors to influence real income independently and interactively. A simple Cobb–Douglas specification is attractive in both respects. Hence we specify as an initial hypothesis

$$y_t = \gamma_0 k_t^{\gamma_1} E_t^{\gamma_2} \qquad \gamma_0, \gamma_1, \gamma_2 > 0 \qquad (4)$$

Greater physical capital (energy) per capita stimulates higher real income per head since $\partial y_t/\partial k_t$ and $\partial y_t/\partial E_t > 0$. Likewise, greater energy consumption per capita raises the effectiveness of capital, and a larger capital stock per head enhances the marginal contribution of greater energy consumption ($\partial^2 y_t/\partial k_t \partial E_t = \partial^2 y_t/\partial E_t \partial k_t > 0$). For statistical purposes we take the logarithm of Equation (4), append a disturbance, θ_t, and estimate

$$\ln y_t = \ln \gamma_0 + \gamma_1 \ln k_t + \gamma_2 \ln E_t + \theta_t \qquad (5)$$

If θ_t follows a first-order process $\theta_t = \xi \theta_{t-1} + \phi_t$, where $\phi_t \sim NID(0, \sigma_\phi^2)$ and $E(\phi_t \theta_{t-1}) = 0$ maximum likelihood estimates are obtained by searching a grid over $|\xi| < 1$ to find the *MLE* of ξ.

IV. STATISTICAL MATTERS

A. Simultaneity

As shown by Equations (3) and (5), E_t and y_t are simultaneously determined. Further, in individual country time-series models, y_t and k_t are both endogenous. Thus instrumental variables are employed to obtain consistent estimates of the coefficients of Equations (3) and (5). For each country I assume a set of four exogenous variables: population, a time trend, the real cost of capital goods, and the real price of energy. Population is certainly exogenous in advanced, Western societies unafflicted by subsistence living standards. The real price of capital for these countries is chiefly determined in internationally competitive markets. Likewise, the base cost of energy is determined in globally competitive markets; domestic prices to end users are then modified by national tax policies, which we regard as exogenous.

B. Autocorrelation

The regression residuals obtained from the initial two-stage least-squares estimates displayed autocorrelation. Accordingly, we assume that the disturbances in Equations (3) and (5) follow first-order autoregressive schemes, $\varepsilon_t = \rho\varepsilon_{t-1} + \mu_t$ and $\theta_t = \xi\theta_{t-1} + \phi_t$, and searched over the values of $|\rho| < 1$ and $|\xi| < 1$ for the maximum likelihood estimates. These values are then employed to transform the endogenous, exogenous, and instrumental variables for structural estimation by three-stage least-squares.

C. Contemporaneous Disturbances

One should expect the disturbance terms in Equations (3) and (5) to be contemporaneously correlated, since exogenous shocks would likely affect both equations simultaneously. If so, a gain in asymptotic efficiency is achieved by estimating the equations simultaneously by three-stage least-squares. In fact, the disturbances of Equations (3) and (5) are positively correlated in each country; hence three-stage least-squares yields full information maximum likelihood estimates.

V. DATA

Data are taken from several sources. They are assembled in an appendix, which is available upon written request. But a brief discussion here will assist in interpreting the statistical estimates. The endogenous variables were discussed in Section II. The exogenous variables are as follows:

1. Population estimates for each country are official United Nations estimates reported in Summers and Heston (1984).
2. Real cost of capital goods is a price index for gross domestic investment, specific to each country and adjusted annually for changes in purchasing power parity, reported in Summers and Heston (1984). This series is a measure of each country's real unit cost of investment goods relative to the cost of capital goods in the United States.
3. Real prices of energy are aggregate country-specific indexes of prices paid by final users, converted to a 1972 U.S. index of 100. Each country's aggregate index is a consumption-share weighted average of constant-cost energy prices (including taxes) in the residential-commercial, transportation, and industrial sectors. The index in each country takes account of changes over time in the fuel consumption mix of each sector, and changes in the relative importance of each sector. The energy price index for each

country is converted to a 1972 U.S. index using purchasing power parity ratios. These individual country indexes are comparable across countries and over time. Observations for the years 1955–1975 are obtained from Dunkerley, Alterman, and Schanz (1980); those for the period 1976–1980 from OECD (1980, p. 52).

VI. PARAMETER ESTIMATES

Estimates of the slope coefficients of Equation (3) are shown in Table 2. By incorporating the a priori restrictions $\beta_1 \leq 0$, $\beta_2 \geq 0$, and $0 \leq \beta_3 \leq 1$, one may efficiently employ one-tail tests of significance. Consider first the lagged adjustment coefficient, β_3. Somewhat surprisingly, Sweden and Canada fail to exhibit significant lags, while France, Germany, and the United States display substantial lags in adjustment. The estimates of β_3 in these three countries imply several years are required to reach equilibrium. For example, the estimate of $\delta = (1-\beta_3) = .44$ for the United States implies a 44 percent adjustment after one year, $.44 + .44(.56) = 69$ percent after two years, $.69 + .44(.31) = 83$ percent after three years, and so on. The speeds of adjustment in France and Germany are quite similar to that in the United States.

Table 2. Energy Demand Equations for Five OECD Countries, 1955–1980

	$\hat{\beta}_1$	$\hat{\beta}_2$	$\hat{\beta}_3$
France	−.100	.310*	.590
(1955–1980)	(.072)	(.099)	(.123)
	$t = -1.38$	$t = 3.14$	$t = 4.80$
Germany	−.159**	.429*	.479*
(1959–1980)	(.069)	(.130)	(.146)
	$t = -2.32$	$t = 3.30$	$t = 3.29$
Sweden	−.389*	1.359*	−.157
(1962–1980)	(.108)	(.327)	(.186)
	$t = -3.61$	$t = 4.16$	$t = -0.84$
Canada	−.353*	.985*	.132
(1955–1980)	(.057)	(.168)	(.153)
	$t = -6.17$	$t = 5.87$	$t = 0.86$
United States	−.192	.465*	.563*
	(.027)	(.082)	(.078)
	$t = -7.15$	$t = 5.64$	$t = 7.23$

Notes: All equations are estimated by three-stage least-squares with adjustment for autocorrelated disturbances. The tabulated estimates are those for the maximum likelihood value of ρ.

The reported *t*-statistics are for tests of the null hypothesis that the parameter is zero against a nonzero alternative.

*Denotes significance at $p \leq .01$; **denotes significance at $p \leq .05$.

Table 3. Long-Run Price and Income Elasticities of Demand in Five OECD Countries

	Price Elasticities	Income Elasticities
France	−.244	.756
	(.176)	(.240)
	$t = -1.39$	$t = 3.15$
Germany	−.307	.823
	(.132)	(.245)
	$t = -2.33$	$t = 3.36$
Sweden	−.389	1.359
	(.108)	(.327)
	$t = -3.61$	$t = 4.16$
Canada	−.353	.985
	(.057)	(.168)
	$t = -6.17$	$t = 5.87$
United States	−.439	1.064
	(.061)	(.187)
	$t = -7.20$	$t = 5.69$

Notes: β_1^* and β_2^* are obtained from three-stage least-squares estimates of Equation (3). Sample periods are 1955–1980 for France, Canada, and United States; 1959–1980 for Germany and 1962–1980 for Sweden.

Consider next the short-run price and income elasticities. All countries except France exhibit highly significant short-run price elasticities (the estimate of β_1 for France is significant at $P \leq .10$); not surprisingly all estimates are quite inelastic. The income elasticities are estimated with high precision, all being significant at $P \leq .01$. France, Germany, and the United States display short-run income elasticities less than one. On the other hand, the short- and long-run elasticities for Sweden and Canada cannot be distinguished since neither country has a statistically significant lagged adjustment. In neither country is the income elasticity different from unity.

The estimated *long-run* price and income elasticities are shown in Table 3. Interestingly enough, all long-run price elasticities fall in the range −.24 to −.44;[9] and all estimates except that for France are highly significant.[10] (Again, the estimated elasticity for France is significant at $P \leq .10$). The income elasticities are estimated with good precision, all being significant at $P \leq .01$. Remarkably, not a single income elasticity differs significantly from unity.

Estimates of Equation (5) are shown in Table 4. Note first that the maximum likelihood search for $|\xi| < 1$ effectively corrects for autocorrelated residuals: The Durbin–Watson statistic permits one to reject the

Table 4. Estimated Coefficients of Capital Per Capita (γ_1) and Energy Per Capita (γ_2)

	$\hat{\gamma}_1$	$\hat{\gamma}_2$	Durbin–Watson d
France	.215**	.630*	1.73
(1955–1980)	(.124)	(.152)	
	t = 1.73	t = 4.15	
Germany	.515*	.486*	1.68
(1959–1980)	(.092)	(.161)	
	t = 5.58	t = 3.02	
Sweden	.468*	.236**	1.46
(1962–1980)	(.060)	(.114)	
	t = 7.77	t = 2.07	
Canada	.540*	.447*	1.99
(1955–1980)	(.108)	(.105)	
	t = 5.02	t = 4.26	
United States	.589*	.474*	1.73
(1955–1980)	(.109)	(.131)	
	t = 5.39	t = 3.62	

Notes: Equations are estimated by three-stage least-squares with adjustment for autocorrelated disturbances. The tabulated values are those associated with the maximum likelihood estimate of the autocorrelation parameter.

The *t*-statistics are for tests of the null hypothesis that the parameter is zero.

*Denotes significance at $p \leq .01$; **denotes significance at $p \leq .05$.

hypothesis of autocorrelation at $P \leq .05$ except in Sweden, for which d lies in the inconclusive region. More importantly, all of the capital and energy coefficients are significant at $P \leq .05$, with 8 being significant at $P \leq .01$. It is clear that both physical capital and energy play major roles in the growth of real incomes. The comparative values of $\hat{\gamma}_1$ and $\hat{\gamma}_2$ suggest that energy holds a particularly prominent role in France, a rather less important one in Sweden, and roughly equal footing for capital and energy in Germany, Canada, and the United States.

VII. THE CONTRIBUTIONS OF CAPITAL AND ENERGY TO GROWTH

Equation (5) can be used to analyze the long-term consequences of capital accumulation and energy consumption. The total rate of per capita income growth is of course

$$d\ln y/dt = \gamma_1(d\ln k/dt) + \gamma_2(d\ln E/dt) \tag{6}$$

Table 5. Actual and Predicted Growth of Real GDP per Capita, and Conditional Growth with Constant Energy per Capita (In Percent)

	France	Germany	Sweden	Canada	United States
1. Actual Annual Growth Rate	3.85	3.44	2.23	2.67	1.77
2. Predicted Annual Growth Rate	3.32	3.48	2.02	2.99	1.91
3. Predicted Annual Growth Rate with Constant Energy per Capita	0.80	2.23	1.53	1.77	1.29

and the conditional growth rate if per capita energy consumption were fixed is

$$(d\ln y/dt)_{E=\text{constant}} = \gamma_1 (d\ln k/dt) \qquad (7)$$

Two questions arise immediately. First, how closely do the predicted growth rates, based on the estimates $\hat{\gamma}_1$ and $\hat{\gamma}_2$, accord with the various countries' actual growth rates? Second, what would each country's predicted growth rate have been if energy consumption had been restricted to initial-period rates? The answer to the first question provides some idea of the usefulness of the estimates; that to the second is an approximation to the role of capital accumulation (and embodied technical progress) in driving real per capita income.

Actual annual growth rates of real GDP per capita for the full sample periods are shown in the first row of Table 5. The predicted growth rates, based on the insertion of $\hat{\gamma}_1$ and $\hat{\gamma}_2$, for each country in Equation (6), are shown in the second row. The predictions are pretty well on the mark, thus suggesting that $\hat{\gamma}_1$ and $\hat{\gamma}_2$, are reliable estimators.

The predicted growth rates, conditional on fixed energy consumption per capita, appear in row 3. If energy consumption had been thus restricted, real income growth would have been very sharply reduced in France. But major retardation would have occurred in Germany, Sweden, and Canada as well. Interestingly enough, the U.S. growth rate would have been least affected because of the relatively large U.S. capital coefficient ($\hat{\gamma}_1$), and the fact that average U.S. growth in energy consumption (1955–1980) was only 1.3 percent. France aside, one may compare the estimates in row 3 to those in row 1 and conclude that higher energy consumption contributed approximately one-third and capital accumulation contributed two-thirds to the actual growth in real per capita incomes.

VIII. CONCLUSIONS

First, it is commonplace in much economics literature to minimize the importance of energy as a determinant of real economic growth. By contrast, many physical scientists stress the role of energy and neglect the independent contribution of real capital formation. The major finding here is that both energy and capital deepening are vital determinants of real income growth.

Second, a simple, two-equation model is used to estimate the per capita demand for energy and the contributions of increased energy and capital per head to real income growth in five OECD countries. Short-run price elasticities of demand for energy are statistically significant but quite inelastic in all countries. Long-run elasticities are also well below unity. Short-run income elasticities are substantially less than one for all countries except Sweden and Canada. Long-run income elasticities are not significantly different from one in any country.

Third, capital and energy consumption per head increased at essentially equal rates in all countries until 1973. Then following the energy price shock of 1973–1974, and continuing rise in real energy prices, per capita energy consumption declined in all countries except Canada. Thenceforth, real incomes increased at much diminished rates because of capital deepening.

The model could be extended in several ways. One would be to specify a vintage capital model. After all, capital stocks of various vintages are designed for pretty specific energy requirements, which in turn depend on expected costs of capital, labor, and energy. To design and estimate a vintage-type macro model for several countries would be a formidable business, yet the payoff could be worthwhile.

Another idea is to experiment with different functional forms in Equations (3) and (5). The choice of log-linear functions was based in part on the comparative ease in estimating them simultaneously. After the fact, however, the logarithmic forms yield economically plausible estimates of the demand and income growth coefficients. I doubt whether different functional forms that permit distinctive roles for capital and energy in Equation (5) and for real energy prices and real income in Equation (3) would much alter the basic conclusions.

Finally, the analysis could usefully be extended to the period beyond 1981, when real energy prices began to drop substantially. The key constraints at present are two: (1) real capital stock estimates for selected OECD countries are available only through 1980; and (2) methodologically consistent real GDP estimates are available through 1985 (Summers and Heston, 1988). The continuing efforts by the OECD Department of Economics and Statistics and ongoing work in the International Comparison Project should alleviate these constraints in the next two or three years.

ACKNOWLEDGMENTS

I am pleased to acknowledge the diligent research assistance of messrs. Philip A. Trostel and Benjamin Contreras. Helpful suggestions were made by participants in the Energy Policy Studies Workshop at M.I.T., particularly Ernst R. Berndt and David O. Wood. Any errors or ambiguities that may remain are my responsibility.

NOTES

1. Economists who have emphasized the role of energy as well as physical capital include Hudson and Jorgenson (1974), Jorgenson (1978, 1984), and Rasche and Tatom (1981).

2. Among economists, Nicholas Georgescu-Roegen has stressed most forcibly the importance of materials and energy, and of the second law of thermodynamics (the entropy law), as essential determinants of long-term real incomes. See Georgescu-Roegen (1971, 1975, 1976, 1979, 1982).

3. See, for example, the time-series studies of the demands for capital, labor, energy, and materials in U.S. manufacturing by Berndt and Wood (1975, 1982), Berndt and Khaled (1979), and in U.S. and Japanese manufacturing by Norsworthy and Malmquist (1983). These studies are based on translog cost functions. The Berndt-Wood (1975), Berndt-Khaled (1979), and Norsworthy-Malmquist (1983) studies found partial substitution elasticities between physical capital and energy to be negative. In a multifactor framework, however, a negative partial substitution elasticity between inputs i and j does not imply fixed proportions between this pair of inputs. For analysis of multifactor derived demand see Moroney and Trapani (1981a, 1981b).

4. Griffin and Gregory (1976) and Griffin (1979) applied a three-input translog cost model (capital, labor, and energy) to manufacturing in nine OECD countries. They found significant, positive partial elasticities of substitution between capital and energy.

5. Berndt and Wood (1987) have recently analyzed the effects of higher energy prices on capital utilization and multifactor productivity growth in U.S. two-digit manufacturing product groups.

6. Summers and Heston (1984) report that these series on real gross domestic product per capita, RGDP*, are based on 1975 international prices adjusted to take account of changes in each country's terms of trade over time. France, West Germany and the United States are among the benchmark countries of the International Comparison Project (ICP) for 1975. Real per capita GDP estimates in 1980 prices have most

recently been reported by Summers and Heston (1988). For the countries analyzed in this paper, the real per capita GDP estimates are not revised by Summers and Heston, but are instead simply restated in terms of 1980 prices. The methodology used in the ICP is discussed in detail in Kravis, Heston, and Summers (1978).

7. A summary of the coefficients used to convert the various forms of energy to standard coal equivalents of heat content is found in the *1981 Yearbook of World Energy Statistics* (United Nations 1983, pp. xxiii–xxvi).

8. This type of partial adjustment model was introduced by Koyck (1954) and has been used widely in applications characterized by short-run adjustment costs. Other partial adjustment processes come to mind. I have not experimented with any others since the Koyck-type model seems theoretically well suited to this task.

9. This range is strikingly similar to that of –.26 to –.5 reported in the survey of earlier studies by Koopmans (1979). It is also consistent with most of the results in the summaries reported by Bohi (1981) and Mittelstadt (1985).

10. Consistent estimates of the asymptotic variances of the long-run elasticities are obtained in the usual way:

$$\text{var}(\beta_1^*) = \text{var}[\beta_1/(1-\beta_3)]$$
$$= [1/(1-\beta_3)]^2 \text{var}(\beta_1) + [\beta_1^2/(1-\beta_3)^4]\text{var}(\beta_3)$$
$$+ [2\beta_1/(1-\beta_3)^3]\text{cov}(\beta_1, \beta_3).$$
$$\text{var}(\beta_2^*) = \text{var}[\beta_2/(1-\beta_3)]$$
$$= [1/(1-\beta_3)]^2 \text{var}(\beta_2) + [\beta_2^2/(1-\beta_3)^4]\text{var}(\beta_3)$$
$$+ [2\beta_2/(1-\beta_3)^3]\text{cov}(\beta_2, \beta_3).$$

REFERENCES

Berndt, E.R., and D.O. Wood. 1975. "Technology, Prices, and the Derived Demand for Energy." *Review of Economics and Statistics*, 57 (August):259–268.

Berndt, E.R., and D.O. Wood. 1982."The Specification and Measurement of Technical Change in U.S. Manufacturing." Pp. 199–221 in *Formal Energy and Resource Models*, edited by J.R. Moroney. Greenwich, CT: JAI Press.

Berndt, E.R., and D.O. Wood. 1987. "Interindustry Differences in the Effects of Energy Price-Induced Capital Utilization Changes on Multifactor Productivity Measurement," In *Advances in the Economics of Energy and Resources*, edited by J.R. Moroney. Greenwich, CT: JAI Press.

Berndt, E.R., and M.S. Khaled. 1979. "Parametric Productivity Measurement and Choice among Flexible Functional Forms." *Journal of Political Economy*, 87 (December): 1220–1245.

Bockris, J.O'M. 1980. *Energy Options: Real Economics and the Solar-Hydrogen System*. New York: Halsted Press.

Bohi, D.R. 1981. *Analyzing Demand Behavior: A Study of Energy Elasticities*. Baltimore, MD: The Johns Hopkins University Press.

Cleveland, C.J., R. Costanza, C.A.S. Hall, and R. Kaufman. 1984. "Energy and the U.S. Economy: A Biophysical Perspective," *Science*, 225 (August): 890–897.

Cook, E. "The Flow of Energy in an Industrial Society." *Scientific American*, 225, 3.

Dunkerley, J.; J. Alterman, and J.J. Schanz Jr. 1980. *Trends in Energy in Industrial Societies,* Palo Alto, CA: Electric Power Research Institute.
Georgescu–Roegen, N. 1971. *The Entropy Law and the Economic Process* Cambridge, MA: Harvard University Press.
Georgescu-Roegen, N. 1975. "Energy and Economic Myths." *Southern Economic Journal,* 41 (January): 347-382.
Georgescu-Roegen, N. 1976. *"Energy and Economic Myths: Institutional and Analytical Economic Essays* . New York: Pergamon Press.
Georgescu-Roegen, N. 1979. "Energy Analysis and Economic Valuation." *Southern Economic Journal,* 45(April): 1023–1058.
Georgescu-Roegen, N. 1982. "Energetic Dogma, Energetic Economics, and Viable Technologies", In *Formal Energy and Resource Models,* edited by J.R. Moroney. Greenwich, CT: JAI Press.
Goeller, H.E., and A.M. Weinberg. 1976. "The Age of Substitutability." *Science,* 191(February): 683–689.
Griffin, J.M. 1979. *Energy Conservation in the OECD: 1980–2000,* Cambridge, MA: Ballinger.
Griffin, J.M., and P.R. Gregory. 1976. "An Intercounty Translog Model of Energy Substitution Responses." *American Economic Review,* 66 (December): 845–857.
Hudson, E.A., and D.W. Jorgenson. 1974. "U.S. Energy Policy and Economic Growth, 1975–2000." *Bell Journal of Economics and Management Science,* 5(Autumn): 461–514.
Jorgenson, D.W. 1978. "The Role of Energy in the U.S. Economy." *National Tax Journal,* 31(September).
Jorgenson, D. W. 1984. "The Role of Energy in Productivity Growth." *The Energy Journal,* 5 (July).
Kmenta, J. 1986. *Elements of Econometrics,* 2nd ed. New York: Macmillan.
Koopmans, T. C. 1979. "Alternative Futures With or Without Constraints on the Energy Technology Mix." In *Directions in Energy Policy,* edited by B.K. Kursunoglu and A. Perlmutter. Cambridge, MA: Ballinger.
Koyck, L.M 1954. *Distributed Lags and Investment Analysis.* Amsterdam: North–Holland.
Kravis, I.B., A. Heston, and R. Summers. 1978. "Real GDP Per Capita for More than One Hundred Countries." *Economic Journal,* 88(June): 215–242.
Kravis, I.B., A. Heston, and R. Summers. 1982. *World Product and Income,* Baltimore, MD: The Johns Hopkins University Press.
Mittelstadt, A. 1985. "Energy Prices: Trends and Prospects." Organization for Economic Cooperation and Development Working Paper Number 22.
Moroney, J.R., and J.M. Trapani. 1981a. "Factor Demand and Substitution in Mineral-Intensive Industries." *Bell Journal of Economics,* 12(Spring):272–284.
Moroney, J.R., and J. M. Trapani. 1981b. "Alternative Models of Substitution and Technical Change in Natural Resource Intensive Industries." In *Modeling and Measuring Natural Resource Substitution,* edited by E. Berndt and B. Field. Cambridge, MA: MIT Press.
Norsworthy, J.R., and D.H. Malmquist. 1983. "Input Measurement and Productivity Growth in Japanese and U.S. Manufacturing." *American Economic Review,* 73(December): 947–967.

Organization for Economic Cooperation and Development. 1983. *Flows and Stocks of Fixed Capital, 1955–1980.* Paris: OECD.

Organization for Economic Cooperation and Development. 1980. *Economic Outlook,* 28(December). Paris: OECD.

Rasche, R.H., and J.A. Tatom. 1981. "Energy Price Shocks, Aggregate Supply, and Monetary Policy: The Theory and the International Evidence." *Carnegie-Rochester Conference Series on Public Policy,* 14(Spring).

Summers, R., and A. Heston. 1984. "Improved International Comparisons of Real Product and Its Composition." *Review of Income and Wealth,* Series 30, 2(June): 207–262.

Summers, R., and A. Heston. 1988. "A New Set of International Comparisons of Real Product and Prices: Estimates for 130 Countries, 1950–1985." *Review of Income and Wealth,* Series 34, Number 1(March): 1–27.

United Nations. 1983. *Yearbook of World Energy Statistics, 1981* New York: United Nations Publishing Service.

Ward, M. 1976. *The Measurement of Capital.* Paris: OECD.

Zellner, A., and M. Geisel. 1970. "Analysis of Distributed Lag Models with Applications to Consumption Function Estimation." *Econometrica,* 38(November): 865–888.

ENERGY CONSUMPTION, CAPITAL, AND REAL OUTPUT:
A COMPARISON OF MARKET AND PLANNED ECONOMIES

John R. Moroney

ABSTRACT

This paper investigates the energy intensiveness and capital intensiveness of gross domestic products for 17 European market economies and 7 East European economies for the years 1978, 1979, and 1980. Gross domestic products and real capital stocks of all countries are expressed in U.S. $1975. Energy consumption is likewise measured in internationally comparable units. The Eastern Bloc consumed on average nearly twice as much energy relative to gross domestic product and relative to capital as the market economies. Despite this, aggregate output per worker in the Eastern Bloc averaged only 60 percent of that in the market economies.

Energy, Growth, and the Environment
Advances in the Economics of Energy and Resources,
Volume 7, pages 205–226.
Copyright © 1992 by JAI Press Inc.
All rights of reproduction in any form reserved.
ISBN: 0-89232-947-5

I. INTRODUCTION

Since World War II, both the centrally planned economies of East Europe and the market economies of West Europe and North America have enjoyed notable growth in real per capita incomes. This growth, in turn, has been driven both in the East and the West by increases in physical capital and energy consumption per worker, together with technological change. Yet the individual roles of capital and energy in the planned and market economies have not been analyzed. Such an analysis is the focal point of this paper. Specifically, we link real gross domestic product (GDP) per worker to physical capital and energy consumption per worker for 7 East European economies, the 7 European members of the Council for Mutual Economic Assistance, and 17 West European countries for the years 1978, 1979, and 1980. Analyzing economy-wide aggregates, we show that energy per dollar of real GDP and energy per unit of capital are markedly higher in East Europe. On the other hand, the average value of capital per worker is twice as high in the West European economies as in those of East Europe.

The finding of higher energy intensity in East Europe is really nothing new. In a comparison of the 7 European CMEA countries and Yugoslavia with 18 OECD countries, around 1980, Slama found the East European economies consumed more than twice as much energy per unit of GDP as the Western countries (Slama 1986, pp. 286–287). Likewise, Gomulka and Rostowski (1988) found that aggregate energy consumption per unit of value added in the industrial sectors of the 7 European CMEA countries ranged from 40 percent to 134 percent higher than that in 15 OECD countries for the years 1962, 1975, and 1982.[1] The Gomulka-Rostowski results are informative for this reason: The higher economy-wide energy intensity among the CMEA economies cannot be attributed chiefly to their comparatively high ratios of industrial output to GDP.

The main novelty here is to analyze simultaneously energy and physical capital as aggregate inputs. The construction of physical capital stocks that are reasonably comparable across market and planned economies permits us to analyze two heretofore unanswered questions:

1. Do the countries of Eastern Europe employ more energy-intensive capital stocks than those of Western Europe?
2. What are the comparative roles of energy and capital as engines

of growth in the two groups of countries? In particular, does the structural relationship between real GDP, capital, and energy differ in a significant way between the two groups?

This paper has two major objectives. The first is to analyze a model of real GDP, energy consumption, and reproducible capital per worker for 17 European market economies and 7 centrally-planned economies: the Soviet Union and the 6 other European members of CMEA. The second is to determine whether the structural relationships in the model differ systematically between the two groups.

An analysis of the relationship among real GDP, capital, and energy per worker is intrinsically useful. For if energy were to exert a negligible influence on real GDP, independently of capital, the economic consequences of diminished energy supplies would be quite different from those if energy were significant in its own right. International cross-section studies by Cook (1971), Darmstadter, Dunkerley, and Alterman (1977), Bockris (1980), and Moroney (1989) established a strong simple correlation between real GDP per capita and energy consumption per capita. Based on scatter diagrams, Cook and Bockris suggested the relationship may be linear, while Moroney found a double-log regression provided a much better statistical fit.[2] These studies all overstate the net association between real GDP per capita and energy per capita, however, because capital stocks were not available. I am able here to incorporate real capital stocks per worker, in U.S. $1975, based on benchmark estimates for the market economies by Leamer (1984) and on official government estimates of constant-price stocks for the CMEA countries. A detailed discussion of the sources and methods used to convert the CMEA capital stocks to U.S. $1975 is provided in the Methodological Appendix.

It is important, as well, to determine if the response of real GDP per worker to energy and capital per worker differs between centrally-planned and market-type economies. The planned economies place far greater reliance on direct resource allocation and less on prices than do market systems. Thus in principle they could use energy and capital either more or less intensively than market economies. This paper shows that the European CMEA countries consume far more energy relative to real GDP and relative to capital than the European market economies.

This paper is organized as follows. Section II presents a snapshot of worldwide energy production and consumption, classified by geopolitical area, for the year 1980. In Section III we discuss a simple model to analyze the relationship between real GDP, energy consumption, and real capital stocks per worker.[3] In Section IV we discuss the international cross-section samples for the years 1978, 1979, and 1980, and present the statistical results. Section V summarizes and concludes.

II. WORLDWIDE ENERGY PRODUCTION AND CONSUMPTION, 1980

The world's production and consumption of primary energy in all forms, including all types of coal, peat, oil shale, crude petroleum, natural gas, natural gas liquids, and electricity is shown in Table 1. All forms of energy are converted to standard coal equivalents, 7,000 calories per gram, according to their specific heat content. Several points are noteworthy.

Table 1. Primary Energy Production and Consumption (Million Metric Tons of Standard Coal Equivalent), 1980

	Production[a]	Consumption[b]
World	9.292	8.566
Africa	.586	.199
North America	2.595	2.850
South America	.343	.249
Asia (excluding China)	1.831	.966
China	.614	.581
Europe (excluding CMEA)	.813	1.560
CMEA (excluding USSR)	.450	.584
USSR	1.938	1.474
Other	.121	.103

Notes: All forms of energy produced and consumed are converted according to their heat content to a standard coal equivalent of 7,000 calories per gram.

[a]Primary energy production includes anthracite and bituminous coal, coke, lignite and brown coal, peat, oil shale, petroleum, natural gas liquids, natural gas, and electricity, including that from hydro, nuclear, and geothermal sources.

[b]Primary energy consumption equals primary energy production plus imports minus exports minus bunker fuels used by ships and aircraft in international transportation minus inventory stock changes. Thus the 7.8 percent difference between world production (9.292 million metric tons) and world consumption (8.566 million metric tons) is attributable to bunker fuel consumption and net worldwide inventory changes.

Source: The conversion factors for the various forms of energy, including adjustments for international differences in conversion factors for various specific fuels, are published in the *1981 Yearbook of World Energy Statistics* (New York: United Nations Publishing Service, 1983).

1. China, the CMEA countries, and the USSR produce approximately one-third of the world's energy. These countries collectively account for roughly 31 percent of aggregate energy consumption. China's production and consumption are practically identical, the CMEA countries are net importers, and the Soviet Union is a major exporter.[4]
2. Europe, excluding the CMEA countries, produces only 8.5%, but consumes more than 18 %, of the world's energy. The European market economies analyzed in this paper account for the lion's share of energy imports required to maintain this level of consumption.
3. North America consumes considerably more than it produces, while Asia and Africa produce far more than they consume.

The energy surplus of the USSR shall be increasingly difficult to maintain. Its disposition bears major economic and political ramifications. First, Soviet oil production is a mature industry that now exhibits rapidly rising costs of extraction.[5] Yet it is essential that the Soviets continue to supply much-needed oil to the CMEA countries for at least the next few years. But to meet domestic and export requirements will be far more costly in the years to come.[6] Most authorities suggest that further development of Soviet energy, chiefly natural gas, for export and domestic consumption will require at least 50 percent greater investment per unit of energy than during the period 1977–1981.[7]

Second, since 1978 the Soviet Union has relied on oil exports to the hard-currency countries of Western Europe as a major source of foreign exchange. After 1973, Soviet exports to Western Europe were at prices essentially on par with Mideast Light Crude. Those to the CMEA were priced at discounts up to 50 percent, and in some years less than one half the world market price (Hewett, 1984, pp. 155, 163). Thus the Soviets face difficult trade-offs in apportioning oil among domestic needs, exports to CMEA countries, and exports to the West.[8] These difficulties are, of course, magnified by rapidly rising extraction costs and the decline in international petroleum prices since 1982.

The nations of Western Europe face a quite different energy problem. Apart from the Netherlands, Norway and the United Kingdom, the European market economies must continue to import hydrocarbons as an essential source of energy. Thus they are particularly vulnerable to interruption in supply.

III. A MODEL OF OUTPUT, ENERGY, AND CAPITAL PER WORKER

An analysis of international differences in labor productivity would ideally include capital, labor, energy, and material inputs. Given internationally comparable input prices, one could estimate the relationships among these inputs using a flexible functional form such as a translog cost function. Such estimates have been obtained with the use of time series data in United States manufacturing by Berndt and Wood (1975, 1982), Berndt and Khaled (1979), and in American and Japanese manufacturing by Norsworthy and Malmquist (1983). Because internationally comparable estimates of material input prices are not available, Griffin and Gregory (1976) estimated the substitution relationships among capital, labor, and energy in manufacturing for a subset of nine OECD countries (see also Griffin, 1979).

An attempt to construct internationally comparable costs of capital for the market and planned economies would be most hazardous. Capital costs are determined in the western economies by competitive or quasi-competitive markets. By contrast, they have historically been governed in the East European economies chiefly by administrative fiat. Because the alternative capital costs are intrinsically incomparable, we analyze the production relationships in a more direct fashion.[9]

When analyzing economy-wide aggregates, the assumption of constant returns to scale is virtually compelling. After all, variation in the scale of the production unit, the plant, is a microeconomic matter applicable to a specific industry. Differences in the size of nations is not an analogous measure of variation in scale. In the following analysis, we thus assume constant returns to scale, which enables everything to be written in labor-intensive form.

Let y = real GDP per worker, k = capital per worker, and E = energy consumption per worker. Consider the relation

$$y = f(k,E) \qquad (1)$$

and denote its derivatives by $\partial y/\partial k = f_k$, $\partial^2 y/\partial k^2 = f_{kk}$, $\partial y/\partial E = f_E$, $\partial^2 y/\partial E^2 = f_{EE}$, and $\partial^2 y/\partial k \partial E = f_{kE}$. Economic theory suggests that in the relevant domains of k and E, $f_k > 0$, $f_E > 0$; $f_{kk} < 0$; $f_{kk} < 0$, $f_{EE} < 0$; and $f_{kE} > 0$. Next define the elasticities of y with respect to k and E as $\partial \ln y/\partial \ln k = \beta_k$ and $\partial \ln y/\partial \ln E = \beta_E$. These elasticities are direct measures of the capital and energy intensities of real output; their magnitudes reflect the relative importance of capital and energy per worker as determinants of labor productivity.

In industrial societies energy is consumed mainly in conjunction with capital stocks; the intensity with which energy is used depends on the historical designs of the capital stocks. These designs, in turn, reflect historical energy costs. Thus two questions arise immediately. What are the magnitudes of β_k and β_E? Do they differ between the western market economies on the one hand and the Soviet Bloc economies on the other?

We estimate these elasticities for the two groups of countries by means of an equation that satisfies the economic properties noted above:

$$\ln y = \alpha + \beta_k \ln k + \beta_E \ln E + \varepsilon, \tag{2}$$

where ε is a disturbance term assumed to be normally distributed with zero mean and constant variance. Specifically, we estimate the coefficients of Equation (2) from three cross-section samples (1978, 1979, and 1980) for the market economies, as a group, and separately for the planned economies. We then test for differences in the capital and energy elasticities between the market and the planned economies.

IV. SAMPLE STATISTICS AND PARAMETER ESTIMATES

A. The International Sample

International cross-section samples are probably the best available source of information concerning the influence of energy consumption and capital stocks on real GDP per worker. The range of variation in real GDP per worker is on the order of 4:1, while that in energy consumption per worker is roughly 8:1, both much larger than those observed in time series for any single country. We link real GDP per worker to aggregate energy consumption and net capital stocks per worker using cross-section samples for the years 1978, 1979, and 1980. Real GDP estimates were prepared under Phase III of the United Nations International Comparison Project for three CMEA countries: Hungary, Poland, and Romania. The more extensive analysis in this paper is made possible because of the meticulous estimates of real GDP per capita of the Soviet Union, Bulgaria, Czechoslovakia, and East Germany published by Summers and Heston (1984), and based on International Comparison Project methodology. A thorough appraisal of this work is given by Marer

(1985). These are the most closely comparable estimates of real output per head ever assembled. They are adjusted annually for changes in purchasing power parity and expressed in U.S. $1975 per capita; hence they are directly comparable across countries and time. The methodology used in their construction is described fully in Kravis, Heston, and Summers (1982), Kravis (1984), and Marer (1985).

Energy consumption per worker is measured comprehensively. It includes all energy attributable to hydrocarbons: crude petroleum, refined petroleum products, natural gas liquids, and natural gas; solid fuels: anthracite and bituminous coal, lignite and brown coal, coke, charcoal, fuelwood, and bagasse; and electricity. All forms of energy are converted to kilograms of standard coal equivalency, 7,000 calories per gram, hence they are fully comparable across countries. These data are obtained from the *1981 Yearbook of World Energy Statistics* (United Nations, 1983). The coefficients used to convert the alternative forms of energy to standard coal equivalency are published in this volume.

For all countries, capital is measured by economy-wide net capital stocks, expressed in U.S. $1975. The reader is referred to the Appendix for a discussion of methodology and sources. Suffice it to say here that the West European capital stocks are developed from benchmark estimates published by Leamer (1984). Those of the CMEA economies are based on official government estimates of each country, compiled by the L.W. International Financial Research, Inc. Research Project on National Income in East Central Europe, and kindly furnished to the author by Dr. Thad P. Alton. I then converted these estimates to U.S. $1975 by applying purchasing power parity (PPP) ratios from the International Comparison Project.

Two points concerning the capital estimates should be mentioned here. The first is an accounting matter. The net capital stocks for the CMEA countries are based on the assumption of 15-year asset lives, the same as for the market economies. Many authorities believe that assets are somewhat longer-lived in the CMEA than in most market economies. Thus a potential error of too-rapid depreciation may understate the CMEA capital stocks. Quantitatively, however, this potential error is minor.[10] The second point concerns differences in the quality of capital. The CMEA capital stocks are reported in national currencies, which I then convert to U.S. $1975 by means of PPP ratios. Some East European experts believe these PPP ratios tend to overstate the dollar value of CMEA capital, because the lower quality of East European capital goods is not adequately represented (Marer, 1985; Bergson, 1987). On the other hand, the principal designers of the International Comparison Project contend that the PPP ratios adequately account for such differences in quality (Marer, 1985, pp. 33ff). In any case, the depreciation and capital quality errors are offsetting; and both are likely to be relatively small.

A scatter diagram of real GDP per capita and energy consumption per capita is shown in Figure 1. The range of energy consumption per person is quite similar for the market and planned economies. But the latter tend to lie on a lower plane, indicating that for given standards of living the planned economies, on the average, use energy more intensively. The data also clearly reveal a flatter response of real GDP per capita with respect to energy per capita, especially among the wealthiest market-type economies: real GDP per person differs only slightly among Switzerland, France, Denmark, Sweden, West Germany, Belgium, and Norway, while energy consumption ranges from 3,636 kg coal equivalent per person in Switzerland to 6,378 in Norway.

Real GDP, energy consumption, and capital stocks per worker in each country for the years 1978, 1979, and 1980 are obtained by dividing the respective per capita values by the country's labor force participation (LFP) rate each year. This adjustment is necessary because the LFP rates in East Europe are considerably higher than those in most West European countries. Among the market economies, only Denmark, Finland, Switzerland, the United Kingdom, and West Germany had LFP rates approaching those of East Europe.

Descriptive sample statistics for the two groups of countries appear in Table 2. The range in real GDP per worker is somewhat larger within the market than within the planned economies. The sample means each year are roughly 60 to 66 % higher in the market economies. From the second column one notes that the average ratios of real GDP per unit of energy in the market economies are approximately twice as large as those in the planned economies.[11] This finding is consistent with the economy-wide results reported by Slama (1986), and with the industrial sector results reported by Gomulka and Rostowski (1988). If one makes the reasonable assumption that the group means are cross-sectionally independent, t-tests for differences between means are $t(1978) = 4.80$, $t(1979) = 4.58$, $t(1980) = 5.14$; each is significant at $P \leq .01$.

The last column of Table 2 shows that the planned economies consume nearly twice as much energy per unit of capital as the market economies. This fact is almost certainly attributable to their historically lower energy

Table 2. Real GDP per Worker, Real GDP per Unit of Energy, and Energy Per Unit of Capital, 1978, 1979, and 1980

Year	Real GDP/N[a] Max	Min	Mean	Real GDP/Energy[b] Max	Min	Mean	Energy/Capital[c] Max	Min	Mean
				Seventeen Market Economies					
1978	16537	4342	12291 (776)	2.704	.924	1.506 (.145)	.8312	.2670	.4875 (.0306)
1979	17237	4799	12686 (782)	2.735	.911	1.475 (.147)	.8417	.2608	.5040 (.0316)
1980	17727	4789	12927 (775)	2.819	.981	1.549 (.147)	.7762	.2600	.4720 (.0277)
				Seven Planned Economies					
1978	9970	5029	7739 (669)	1.039	.623	.771 (.0492)	1.595	.5536	.9468 (.1351)
1979	10186	5190	7819 (670)	1.040	.636	.766 (.0487)	1.561	.5249	.9088 (.1320)
1980	10360	5075	7765 (724)	1.031	.618	.748 (.0516)	1.549	.5053	.8802 (.1321)

Notes: The market economies are Austria, Belgium, Denmark, Finland, France, Iceland, Italy, Malta, Netherlands, Norway, Portugal, Spain, Sweden, Switzerland, Turkey, the United Kingdom, and West Germany. The planned economies are Bulgaria, Czechoslovakia, East Germany, Hungary, Poland, Romania, and the Soviet Union.
[a]Real GDP/N is measured in U.S. $1985, adjusted for international differences in purchasing power parity.
Source: Summers and Heston (1984).
[b](Real GDP/Energy) is GDP in U.S. $1975/kilogram of standard coal equivalency.
Sources: Real GDP from Summers and Heston (1984); energy from United Nations (1983).
[c]Energy/Capital is kilograms of standard coal equivalency per dollar of net capital stock (expressed in U.S. $1975).

prices and to their slow response to rising energy costs since 1973. Assuming again that the group means are cross-sectionally independent, t-tests for differences between means are $t(1978) = 3.32$, $t(1979) = 2.98$, and $t(1980) = 3.02$; each is significant at $P \le .01$. The evidence is clear that capital stocks are considerably more energy intensive in the Eastern Bloc than among the nations of Western Europe. We now turn to a formal analysis of per capita real GDP, capital, and energy in the two groups of countries.

B. Parameter Estimates

The disturbance terms in the cross-section regressions are unlikely to be independent across countries or in adjacent years. Accordingly,

Table 3. Parameter Estimates of Equation (2), Market Economies and Planned Economies

Year	$\alpha_{(m)}$	$\beta_{k(m)}$	$\beta_{E(m)}$	R^2
Market Economies				
1978	4.315**	.341**	.191*	.95
	(.360)	(.086)	(.081)	
1979	4.352**	.358**	.170*	.96
	(.268)	(.058)	(.053)	
1980	4.299**	.385**	.149*	.95
	(.308)	(.069)	(.063)	
Planned Economies				
1978	2.718	.351*	.319	.82
	(2.005)	(.164)	(.220)	
1979	2.305	.338*	.377*	.82
	(1.566)	(.120)	(.152)	
1980	1.513	.405**	.391*	.86
	(1.385)	(.098)	(.128)	

Notes: *Denotes significance at $p \le .01$.
**Denotes significance at $p \le .05$.

parameters are estimated using the method suggested by Zellner (1962), which incorporates the covariances of the disturbances within each group of countries. The parameters of Equation (2) are estimated separately for the market and planned economies. Estimates are shown in Table 3. Asymptotic standard errors are listed in parentheses beneath the estimated coefficients. Consider first the individual cross-section estimates for the market economies. Both the capital and energy coefficients are stable and statistically significant in all regressions, but note that the capital coefficients are estimated with higher precision, and are roughly twice as large as the energy coefficients.

Estimates for the planned economies are a different story. The energy coefficients are essentially the same size as the capital coefficients. But more importantly, these energy coefficients are considerably larger than those of the market economies. All capital coefficients are significant at $P \le .05$, but only two of the three energy coefficients meet this test of significance. Although the estimated energy coefficients of the CMEA countries increase from 0.319 in 1978 to 0.391 in 1980, I would hesitate to read much into this, especially since the coefficient in 1978 has such a large asymptotic standard error.

We next combine all observations from both groups of countries, but allow the two groups to have different intercept and slope coefficients. This procedure utilizes all of the sample information in a flexible way. The equation estimated is

$$\ln y_i = \alpha_{(m)} + \delta_\alpha D_i + \beta_{k(m)} \ln k_i + \delta_k D_i \ln k_i + \beta_{E(m)} + \delta_E D_i \ln E_i + \varepsilon_i \quad (3)$$

where $D_i = 0$ if observations are for market economies and $D_i = 1$ for the planned economies. As before, the estimated capital coefficient for market economies is $\beta_{k(m)}$, while that for planned economies is $\beta_{k(m)} + \delta_k$. Likewise, the respective energy coefficients are $\beta_{E(m)}$ and $\beta_{E(m)} + \delta_E$. The estimated equation is

$$\ln y_i = 4.349 - 2.439 \, D_i + .399 \ln k_i - .065 \, D_i \ln k_i$$
$$ (.195) \quad (.861) \quad\ (.050) \quad\ \ (.074)$$

$$+ .127 \ln E_i + .296 \, D_i \ln E_i \, ; R^2 = .95 \quad (3a)$$
$$ (0.48) \quad\ \ (.123)$$

Two features stand out. First, the negative estimate of δ_α is statistically significant at $P \leq .01$; the planned economies use capital and energy in generally less efficient ways than the market systems.[12] Put somewhat differently, the aggregate production function of the planned economies lies on a lower plane than that of the market economies. Second, the positive estimate of δ_E is significant at $P \leq .05$, while the negative estimate of δ_k is not statistically significant. The CMEA countries display a significantly greater energy intensity than the market economies. Yet the estimated capital coefficients do not differ in a statistically significant way.

IV. SUMMARY AND CONCLUSIONS

Increasing consumption of energy and real capital formation have been the mainsprings of economic growth for both market-oriented and centrally-planned economies since World War II. The main contribution of this paper is to link East-West differences in real output per worker to energy consumption and physical capital stocks per worker.

Two conclusions are clear. First, it is indisputable that the CMEA economies as a group consume, on the average, approximately twice as much energy per unit of capital and per unit of real GDP as the economies of Western Europe. This pattern of intensive energy use may well be a rational response by end users to historical energy subsidies and to central planning emphasizing energy-intensive industry.

Second, both the market and the CMEA countries have relied on capital accumulation and increased energy consumption as sources of economic growth. But the CMEA has relied to a greater extent on energy. Equations (2) and (3) reveal their energy coefficients to be much larger than those of the market economies.

These conclusions are based on aggregate Cobb-Douglas production functions with constant returns to scale imposed. Constant returns is not at all restrictive for this sort of economy-wide analysis. The unitary substitution elasticities embedded in the Cobb-Douglas model may seem constraining. But at this level of aggregation I doubt that it makes much difference. Our findings of relatively high East European energy intensity reinforce the earlier economy-wide results of Slama (1986), which were not based on any sort of formal production model. And they are consistent with the findings of Gomulka and Rostowski (1988), based on input-output tables for the industrial sectors of the European CMEA and 15 OECD countries.

METHODOLOGICAL APPENDIX

Estimation of Real Capital Stocks for Market and Nonmarket Economies, 1978, 1979, and 1980

Market Economies

The starting point for all market economies is a set of benchmark net capital stock estimates, in U.S. $1975, by Edward E. Leamer (1984, Appendix B). Leamer's estimated constant-dollar capital stocks are comprehensive, economy-wide measures that include all reproducible capital assets of public and private enterprises, private nonprofit institutions, and net inventory valuation adjustments. To construct the net capital stocks, Leamer used a perpetual inventory method, assuming average asset lives of 15 years. He then applied a double declining balance depreciation rate of 13.3 percent to the net capital stock remaining each year, and added gross investment annually. The capital stocks are converted to U.S. dollars using market exchange rates.

Starting with Leamer's 1975 benchmarks, I obtain annual gross investment in each country for the years 1976–1980, expressed in U.S.

Table A1. Energy, Capital, and Real Output: Real Gross Domestic Product Per Capita (RGDP in U.S. $1975)[a] and Energy Per Capita (Kilograms of Standard Coal Equivalency)[b]

Country	RGDP/Capita 1980	RGDP/Capita 1979	RGDP/Capita 1978	E/Capita 1980	E/Capita 1979	E/Capita 1978
Market Economies						
Austria	6052	5816	5504	4079	4276	3942
Belgium	6293	6138	5999	5973	6448	6133
France	6678	6641	6442	4355	4465	4174
West Germany	6967	6820	6485	5729	6103	5827
Netherlands	5856	5845	5756	5970	6413	6228
Switzerland	6610	6307	6159	3636	3592	3679
United Kingdom	4990	5067	5004	4791	5188	4964
Denmark	6746	6833	6620	5283	5653	5326
Finland	5939	5820	5416	5524	5392	5055
Iceland	5836	5836	5746	4930	5464	4911
Norway	6825	6619	6334	6378	6400	5969
Sweden	7142	6950	6703	5431	5766	5596
Italy	4661	4415	4209	3374	3468	3039
Malta	3476	3279	3030	1266	1199	1197
Portugal	3092	2853	2874	1219	1135	1063
Spain	4264	4233	4187	2355	2394	2154
Turkey	2069	2083	1893	734	792	712
Planned Economies						
Bulgaria	3437	3555	3455	5341	5050	4828
Czechoslovakia	4908	4856	4852	6393	6355	6466
East Germany	5532	5409	5264	7271	7164	7029
Hungary	3861	3856	3843	3745	3709	3697
Poland	3509	3668	3770	5000	4815	4709
Romania	2766	2839	2761	4477	4467	4434
USSR	3943	3891	3866	5551	5522	5377

Sources: [a]Real Gross Domestic Product Per Capita, in U.S. $1975, and adjusted annually for changes in purchasing power parity for each country, are taken from Heston and Summers (1984).
[b]Energy per capita, converted to kilograms of standard coal equivalency (7,000 calories per gram) per year, are obtained from *1981 Yearbook of World Energy Statistics* (United Nations, 1983).

$1975 adjusted for purchasing power parity, from Summers and Heston (1984). Following Leamer's assumptions of 15-year asset lives and double declining balance depreciation of .133 annually, I estimate the net capital stock in each country for each year t ($t = 1976, \ldots, 1980$) as

$$K_t^{net} = .867 K_{t-1}^{net} + I_t^{gross}$$

where all values are expressed in constant U.S. $1975. In this way,

Table A2. Energy, Capital, and Real Output: Real Net Capital Stock (U.S. $1975) Per Capita

Country	Capital/Capita 1980	Capital/Capita 1979	Capital/Capita 1978
Market Economies			
Austria	9476	9228	9044
Belgium	10173	10080	9964
France	11503	11213	10866
West Germany	12173	11694	11172
Netherlands	9602	9541	9392
Switzerland	13987	13772	13778
United Kingdom	6172	6164	5972
Denmark	11260	11379	11009
Finland	10772	10773	10779
Iceland	11726	11430	11209
Norway	13880	13611	13675
Sweden	11927	11816	11970
Italy	6383	5884	5581
Malta	3205	3047	2882
Portugal	3239	2881	2617
Spain	5579	5425	5229
Turkey	1597	1525	1436
Planned Economies			
Bulgaria	3447	3236	3027
Czechoslovakia	8954	8522	8127
East Germany	10116	9703	9300
Hungary	7411	7066	6678
Poland	5104	4888	4607
Romania	4228	3971	3680
USSR	8560	8096	7668

Sources:[a] For market economies the 1975 benchmark capital stocks, in 1975 U.S. dollars, are taken from E.E. Leamer, *Sources of International Comparative Advantage* (Cambridge, MA: MIT Press, 1984). For subsequent years, gross investment, in U.S. $1975, are obtained from Robert Summers and Alan Heston, "Improved International Comparisons of Real Product and Its Composition," *Review of Income and Wealth*, 30, 2 (June 1984) pp. 207–262. The same methodology employed by Leamer concerning average asset lives and depreciation is followed: 15 year asset lives and a double declining balance annual depreciation of 13.3 percent.

[b] For planned economies other than the USSR the constant value capital stocks, in national currency units, are obtained from Thad P. Alton, *Fixed Capital Stock in Eastern Europe, 1960–1984*, Research Project on National Income in East Central Europe, L.W. International Financial Research, Inc. The Soviet capital stocks are obtained from *Narodnoe Khoziastvo SSSR v 1980 godv* (National Economy of the USSR in 1980), which were kindly furnished by Professor Gertrude Schroeder Greenslade of the University of Virginia. Each country's capital stock was converted to U.S. $1975 using purchasing power parity ratios developed in the World Bank International Comparison Project. For details, see the Methodological Appendix.

Table A3. Energy, Capital, and Real Output: Labor Force Participation Rates (Percent)

Country	Participation Rate 1980	Participation Rate 1979	Participation Rate 1978
Market Economies			
Austria	44.6	45.0	45.4
Belgium	39.2	39.5	39.7
France	42.9	43.2	43.4
West Germany	46.6	46.9	47.1
Netherlands	38.8	39.1	39.3
Switzerland	48.9	49.1	49.1
United Kingdom	47.3	47.4	47.6
Denmark	47.6	47.8	48.0
Finland	48.3	48.5	48.6
Iceland	40.7	40.6	41.2
Norway	38.3	38.4	38.5
Sweden	44.4	44.5	44.6
Italy	37.4	37.5	37.6
Malta	35.6	35.5	34.5
Portugal	38.6	38.5	38.5
Spain	35.1	35.1	35.3
Turkey	43.6	43.4	43.2
Planned Economies			
Bulgaria	53.0	52.1	52.7
Czechoslovakia	49.5	49.4	49.5
East Germany	52.8	53.1	53.4
Hungary	47.4	47.3	47.2
Poland	54.0	54.3	54.4
Romania	54.9	54.7	54.5
USSR	49.8	49.9	50.2

Source: Yearbook of Labour Statistics.

methodologically consistent estimates of net book values of capital are obtained for each market economy for the years 1978, 1979, and 1980.

The physical composition of these stocks differs from country to country. One cannot appeal to any neat aggregation theorems to claim strict comparability. Nonetheless, these stocks are based on a consistent, perpetual inventory methodology, and appear to be the most closely comparable estimates available.

Nonmarket (CMEA) Economies

We describe here the methodology used to construct estimates of reproducible capital stocks for each nonmarket country and to express

Energy, Consumption, Capital, and Real Output 221

them in 1975 U.S. dollars. The starting point for each non-Soviet country is constant-price gross capital stocks prepared for the Research Project on National Income in East Central Europe by L.W. International Financial Research, Inc. These estimates were kindly provided to me by the Research Project Director, Dr. Thad P. Alton. The constant-price gross capital stocks for the Soviet Union are official Soviet Government estimates, which were graciously provided by Dr. Gertrude Schroeder Greenslade, Department of Economics, University of Virginia.

Because questions of international comparability are so important, it seems best to discuss in some detail the definitions of the capital stocks and the method of converting them to U.S. $1975.

The published constant-price capital stocks are expressed in terms of national currencies. Several alternatives could be used to convert them to a common, internationally comparable currency. Among the alternatives are official exchange rates, commercial exchange rates, tourist exchange rates for convertible currencies, or a foreign trade currency multiplier (a coefficient that relates the domestic wholesale price directly to the foreign currency price ex post facto). The problems in using any such exchange rates.

Bulgaria. Gross book value of the fixed, aggregate capital stock is reported in constant 1977 replacement values (million leva). These are converted to U.S. $1975 using the PPP ratios published in Marer (1985, p. 42). The gross book values are then reduced to net book values using the assumptions of a 15-year asset life, and a double-declining balance depreciation rate $\delta = .133$.

Czechoslovakia. Gross book value of the fixed, aggregate capital stock is reported in constant 1977 replacement values (million crowns). These are converted to U.S. $1975 according to the PPP ratios published in Marer (1985, p. 44). The gross book values are reduced to net book values using the asset lives and depreciation rates above.

German Democratic Republic. Gross economy-wide capital stocks are reported in constant 1980 full replacement values (million marks). These stocks are converted to U.S. $1975 according to the adjusted PPP ratio published in Marer (1985, p. 7), and reduced to net book values using the asset lives and depreciation rates above.

Hungary. Gross aggregate capital stocks are reported in constant 1976 replacement costs (billion forints). These stocks are converted to

$1975 U.S. using the 1975 Hungarian PPP ratio for capital formation published in. *World Tables* for the World Bank ICP (1983, p. 569). These gross stocks are reduced to net stocks using the asset lives and depreciation assumptions above.

Poland. Gross aggregate capital stocks are reported in constant 1977 replacement values (billion zlotys). They are transformed to U.S. $1975 using the 1975 PPP for capital formation published in *World Tables* for the World Bank ICP (1983, p. 569), and reduced to net stocks using the assumptions above.

Romania. Gross aggregate capital stocks are published in constant 1963 "full inventory" values (billion lei). Between 1963 and 1975 the estimated inflation in capital goods prices was, on average, three percent annually. The gross capital stocks, in 1963 prices, are expressed in terms of 1975 lei and converted to U.S. $1975 according to the 1975 PPP for capital formation published in *World Tables* (World Bank, 1983, p. 569). The gross stocks are transformed to net stocks according to the standard assumptions.

Soviet Union. Gross aggregate capital stocks for the Soviet Union are published in *Narodnoe Khoziaistvo SSSR v 1980 godv* (National Economy of the USSR in 1980) (Moscow, 1981). These estimates are expressed in constant 1973 rubles. They are converted to U.S. $1975 using a weighted average of the PPP for new machinery and equipment, 1976, and the PPP for construction and other investment, 1976. The PPP for new machinery and equipment covers 245 items, that for construction investment includes 277 items. Both PPPs are based on adjustments of all components to maintain comparable qualities across the United States and USSR. The PPPs are published in Edwards, Hughes, and Noren (1979, p. 379). The gross capital stocks are reduced to net stocks in the usual way.

ACKNOWLEDGMENTS

I am pleased to acknowledge the helpful research assistance of Philip A. Trostel, and the constructive comments of Professors Gertrude Schroeder Greenslade, James M. Griffin, J.R. Hanson, and Judith Thornton. These scholars are in no way responsible for any ambiguities or errors that may remain. I am also grateful for the detailed comments of Josef C. Brada and an anonymous referee.

An earlier version of this paper appeared in *The Journal of Comparative Economics*, Volume 14 (June 1990). It is reprinted here with the kind permission of Josef C. Brada, Editor of the *Journal* and of Academic Press.

NOTES

1. In a sample of 25 Western economies and 8 centrally planned economies (including China) for 1972, Hewett (1980) found that the planned economies exhibited relatively high energy consumption in relation to GNP, but the difference was not always statistically significant. Schroeder (1985) reported that Soviet energy consumption per dollar of GNP in 1982 exceeded that in the United States and several countries in the European Economic Community. In a pairwise comparison of Czechoslovakian input-output tables in 1962 and Austrian tables in 1964, Drabek (1988) found the natural resource input use to be quite similar in the two countries. Drabek's comparisons were based both on the direct and direct plus indirect coefficient matrices for all natural resources, but did not focus specifically on the relative use of energy.

2. The later work by Dunkerley, Alterman, and Schanz (1980) also used the double logarithmic specification, in which the log of energy per capita was specified as the dependent variable, and the log of real GDP per capita served as a regressor. They also used an energy price index to attempt to explain international variation in energy consumption. This index was usually insignificant, and occasionally exhibited a positive regression coefficient.

3. Capital services would be a preferable measure of real capital flows. We must rely on net capital stocks because internationally comparable capital services are nowhere available. Indeed, estimation of internationally comparable net capital stocks, expressed in units of constant purchasing power, is a formidable task. If services are proportional to such stocks, then our model yields unbiased statistical estimates and tests of significance.

4. Ebel (1980) estimates that during the late 1970s the Soviet Union supplied more than 66 percent of the oil consumed by Eastern Bloc countries, whereas only about 16 percent of the oil consumed by these countries was imported from non-Soviet sources. The remaining 18 percent of oil consumption was produced chiefly by Romania. A recent study by the U.S. Central Intelligence Agency (1985) reports that by 1980 the Soviet Union supplied more than 90 percent of all oil imports by other CMEA countries, Bulgaria, Czechoslovakia, East Germany, Hungary, Poland, and Romania, but that in 1982 the Soviets began reducing the volumes of concessionary oil delivered to CEMA countries, except Poland, and raised its export prices. Among the non-Soviet members of CMEA, only Romania has been practically self sufficient in producing domestically its oil consumption requirements. For further discussion of Soviet energy subsidies to Eastern Europe, see the excellent study by Marese and Vanous (1983) and the highly informative work of Hewett (1984). The political interests served by Soviet energy subsidies to Eastern Bloc countries are carefully discussed by Hardt (1984) and Schroeder (1985).

5. Hewett (1984, pp. 41 ff.) estimates that Soviet capital costs of extracting net

incremental oil in 1980 was nearly four times as large as the corresponding costs during the years 1966–1970.

6. Scanlan (1982) suggests three major reasons for the continuing dependence of European CMEA countries on imported Soviet oil at least through the early 1990s: (1) these countries import roughly 75 percent of their oil from the Soviet Union; (2) they have an exceptionally high level of energy consumption per capita; and (3) apart from Romania, the CMEA countries appear to have quite limited flexibility in substituting other fuels for oil in their primary energy mix.

7. See Gustafson (1982), Hewett (1982), and Kurtzweg and Tretyakova (1982).

8. In 1986, coal accounted for 25 percent, natural gas for 36 percent, and oil for 39 percent of Soviet primary energy production (*Yearbook of Energy Statistics*, 1989). On the other hand, coal accounted for 31 percent natural gas for 40 percent, and oil for 29 percent of Soviet primary energy consumption. Dienes and Merkin (1985) emphasize that the European CMEA countries have responded to the phaseout of concessionary Soviet oil prices by attempts to increase alternative domestic energy sources and by administrative demand management, energy substitution, and higher energy prices. Yet they suggest that household and wholesale energy prices remain below marginal production costs, thus prohibiting rational energy management. In their view, only East Germany and Hungary among the CMEA countries have adopted reasonably coherent energy policies permitting a moderation of energy consumption relative to Gross Domestic Product.

9. See Bergson (1987) for a similarly direct attempt to analyze differences in productivity in the materials producing sectors of seven market economies and four planned economies. The inputs employed in Bergson's analysis are capital and agricultural land per capita, but energy is omitted.

10. For example, if the true CMEA asset lives were 25 years rather than 15, the correct double declining balance depreciation rate would be 8.0 percent rather than 13.3 percent. Denote the true net capital stock in year t as K_t^* and the incorrectly estimated stock as K_t. The true net capital stock, after gross investment in year t, I_t, is $K_t^* = .92 K_{t-1}^* + I_t$, while the estimated stock is $K_t = .867 K_{t-1} + I_t$. Given the high investment rates among CMEA countries, one may confirm that a difference of up to 10 years between true and assumed asset lives makes only a minor difference between the true and the estimated net capital stocks.

11. Marer (1985, pp. 114–115) has also noted that the CMEA countries tend to be relatively high and perhaps wasteful consumers of energy. They are certainly relatively high consumers; whether wasteful is not clear, because their intensive consumption may be a rational response to historically subsidized energy prices. In a personal communication to the author, Professor Gertrude Schroeder Greenslade expressed her view that firms in centrally planned economies respond to output targets and input allocations determined by planners, and are only weakly sensitive eo changes in relative prices. In her view, the capital stocks in the Soviet Union and other CMEA countries entail wasteful usage of energy in view of post–1979 real worldwide energy prices.

12. Bergson (1987) found in a study of four planned and seven market economies that the former were generally less efficient in their use of capital and agricultural land. Energy was not incorporated in Bergson's analysis.

REFERENCES

Bergson, A. 1987. "Comparative Productivity: USSR, Eastern Europe, and the West." *American Economic Review*, 77, 3: 342–357.

Berndt, E.R., and M.S. Khaled. 1979. "Parametric Productivity Measurement and Choice among Flexible Functional Forms." *Journal of Political Economy*, 87, 6: 1220–1245.

Berndt, E.R., and D.O. Wood. 1975. "Technology, Prices, and the Derived Demand for Energy." *Review of Economies and Statistics*, 57, 3: 259–268.

Berndt, E.R., and D.O. Wood. 1982. "The Specification and Measurement of Technical Change in U.S. Manufacturing." Pp. 199–221 in *Formal Energy and Resource Models*, edited by J.R. Moroney. Greenwich, CT: JAI Press.

Bockris, J.O'M. 1980. *Energy Options: Real Economies and the Solar-Hydrogen System*. New York: Halsted Press.

Cook, E. 1971. "The Flow of Energy in an Industrial Society." *Scientific American* 225, 3.

Darmstadter, J., J. Dunkerley, and J. Alterman. 1977. *How Industrial Societies Use Energy*. Baltimore, MD: The Johns Hopkins University Press for Resources for the Future.

Dienes, L., and V. Merkin. 1985. "Energy Policy and Conservation in Eastern Europe." In *East European Economies: Slow Growth in the 1980's*. Washington, DC: U.S. Government Printing Office.

Drabek, Z. 1988. "The Natural Resource Intensity of Production Technology in Market and Planned Economies: Austria vs. Czechoslovakia. *Journal of Comparative Economics*, 12, 2: 217–227.

Dunkerley, J., J. Alterman, and J.J. Schanz, Jr. 1980. *Trends in Energy Use in Industrial Societies*, Palo Alto, CA: Electric Power Research Institute.

Ebel, R.E. 1980. "Energy Demand in the Soviet Bloc and the People's Republic of China." In *International Energy Strategies*, Cambridge, MA: Oelgeschlager, Gunn, and Hain, Inc.

Edwards, I., M. Hughes, and J. Noren. 1979. "U.S. and U.S.S.R.: Comparisons of GNP." In *Soviet Economy in a Time of Change*. Washington, DC: U.S. Government Printing Office.

Gomulka, S., and J. Rostowski. 1988. "An International Comparison of Material Intensity." *Journal of Comparative Economics*, 12, 4: 475–501.

Griffin, J.M., and P.R. Gregory. 1976. "An Intercountry Translog Model of Energy Substitution Responses." *American Economic Review*, 66, 4: 845–857.

Griffin, J.M. 1979. *Energy Conservation in the OECD: 1980–2000*, Cambridge, MA: Ballinger.

Gustafsen, T. 1982. "Soviet Energy Policy." In *Soviet Economy in the 1980's: Problems and Prospects*. Washington, DC: U.S. Government Printing Office.

Hardt, J.P. 1984. "*Soviet Energy Policy in Eastern Europe*." In Soviet Policy in Eastern Europe, edited by S.M. Terry. New Haven, CT: Yale University Press.

Hewett, E.A. 1980. "Alternative Econometric Approaches for Studying the Link Between Economic Systems and Economic Outcomes." *Journal of Comparative Economics*, 4, 3: 274–294.

Hewett E.A. 1982. "Near-Term Prospects for Soviet Natural Gas Industry, and the Implications for East-West Trade." In *Soviet Economy in the 1980's: Problems and Prospects*. Washington, DC: U.S. Government Printing Office.

Hewett E.A. 1984. *Energy, Economics, and Foreign Policy in the Soviet Union*, Washington, DC: The Brookings Institution.

Kravis, I.B. 1984. "Comparative Studies of National Incomes and Prices." *Journal of Economic Literature*, 22, 1: 1–39.

Kravis, I.B., A. Heston, and R. Summers. 1982. *World Product and Income: International Comparisons of Real Gross Product*. Baltimore, MD: The Johns Hopkins University Press.

Kurtzweg, L., and A. Tretyakova, 1982. "Soviet Energy Consumption: Structure and Future Prospects." In *Soviet Economy in the 1980's: Problems and Prospects*. Washington, DC: U.S. Government Printing Office.

Leamer, E.E. 1984. *Sources of International Comparative Advantage*. Cambridge, MA: MIT Press.

Marer, P. 1985. *Dollar GNPs of the U.S.S.R. and Eastern Europe*. Baltimore, MD: The Johns Hopkins University Press.

Marese, M. and J. Vanous. 1983. *Soviet Subsidization of Trade with Eastern Europe*, Berkeley, CA: Institute of International Studies, University of California.

Moroney, J.R. 1989. "Output and Energy: An International Analysis." *The Energy Journal*, 10, 3: 1–18.

Norsworth, J.R. and D.H. Malmquist. 1983. "Input Measurement and Productivity Growth in Japanese and U.S. Manufacturing." *American Economic Review*, 73, 4: 947–967.

Scanlan, T. 1982. "The Outlook for Soviet Oil." *Science*, 217: 325–330.

Schroeder, G.E. 1985. "The Soviet Economy." *Current History*, 84: 309–342.

Shabad, T. 1983. "The Soviet Potential in Natural Resources: An Overview." In *Soviet Natural Resources in the World Economy*, edited by R.C. Jensen, T. Shabad, and A.W. Wright, Chicago: University of Chicago Press.

Slama, J. 1986. "An International Comparison of Sulphur Dioxide Emissions." *Journal of Comparative Economics*, 10, 3: 277–292.

Summers, R., and A. Heston. 1984. "Improved International Comparisons of Real Product and Its Composition." *Review of Income and Wealth*, 30, 2: 207–262.

United Nations. 1983. *Yearbook of World Energy Statistics, 1981*. New York: United Nations Publishing Service.

U.S. Central Intelligence Agency. 1985. "East European Energy Outlook Through 1990." In *East European Economies: Slow Growth in the 1980's*. Washington, DC: U.S. Government Printing Office.

Van Brabant, J.M. 1985. *Exchange Rates in Eastern Europe*, World Bank Staff Working Papers Number 778. Washington, DC: The International Bank for Reconstruction and Development.

World Tables. 1983. Baltimore, MD: The Johns Hopkins University Press.

Zellner, A. 1962. "An Efficient Method of Estimating Seemingly Unrelated Regressions and Tests for Aggregation Bias." *Journal of the American Statistical Association*, 57, 2: 348–368.

Advances in the Economics of Energy and Resources

Edited by **John R. Moroney,** *Department of Economics, Texas A&M University*

REVIEWS: "The choice of issues to be covered and the general quality of the papers is outstanding, and the volume is highly recommended."

-The Economic Journal

"...the papers represent excellent research in the area of energy and resources economics. The coverage of topics is broad, from macro-economic models to specific market and policy analyses, from theoretical to empirical studies, and from domestic to international in scope. These volumes seem most useful as a collection of readings for use in a graduate seminar or as supplemental readings in a graduate course in natural resources economics. They are also an excellent source material for the professional economist not familiar with research in the area."

- Southern Economic Journal

"...for the researcher in energy economics the articles are valuable steps forward on topics just now coming under research microscopes. Without these and hundreds more building blocks, energy economics can never get out of its infancy."

- Choice

Volume 1, The Structure of Energy Markets
1979, 310 pp. $63.50
ISBN 0-89232-016-8

CONTENTS: Energy-Economy Interactions: The Fable of the Elephant and the Rabbit?, *William W. Hogan and Alan S. Manne, Stanford University.* **Input Prices, Substitutions, and Products Inflation,** *John R. Moroneyy, Tulane University and Alden Toevs, University of Oregon.* **Substitution Among Energy, Capital, and Labor Inputs in U.S. Manufacturing,** *Robert Halvorsen and Jay P. Ford, University of Washington.* **The Stanford Energy/Economic Model,** *Thomas J. Connolly, George B. Danzig and Shailendra C. Parikh, Stanford University.* **New Car Efficiency Standards and the Demand for Gasoline,** *James L. Sweeney, Stanford University.* **Factors Leading to Structural Change in the U.S. Oil-Refining Industry in the Postwar Period,** *Stephen C. Peck, Electric Power Research Institute, Palo Alto, and Scott Harvey, University of California—Berkeley.* **Integration and Innovation in the Energy Markets,** *David J. Teece, Stanford University.* **The Economics of the**

Throwaway Nuclear Fuel Cycle, *John B. Gordon and Martin L. Baughman, University of Texas—Austin.* **Prospects for Nuclear Power in the Developing Countries,** *Alan M. Strout, Massachusetts Institute of Technology.* **Financial Markets and the Adjustment to Higher Oil Prices,** *Tamir Agmon, Tel Aviv University, James L. Paddock and Donald R. Lessard, Massachusetts Institute of Technology.*

Volume 2, The Production and Pricing of Energy Resources
1979, 250 pp. $63.50
ISBN 0-89232-079-6

CONTENTS: Alternative Methods of Oil Supply Forecasting, *M.A. Adelman and Henry D. Jacoby, Massachusetts Institute of Technology.* **A Basin Development Model of Oil Supply,** *Paul L. Eckbo, Massachusetts Institute of Technology.* **Estimating a Policy Model of U.S. Coal Supply,** *Martin B. Zimmerman, Massachusetts Institute of Technology.* **The Rate of Petroleum Exploration and Extraction,** *Russell S. Uhler, University of British Columbia.* **Uncertainty and the Optimal Supply for an Exhaustible Resource,** *Geoffrey Heal, University of Sussex.* **Search Strategies and Private Incentives for Resource Exploration,** *Richard J. Gilbert, University of California—Berkeley.* **Increasing Extraction Costs and Resource Prices,** *Donald A. Hanson, Southern Methodist University.* **Staving Off the Backstop: Dynamic Limit-Pricing with a Kinked Demand Curve,** *Stephen W. Salant, Board of Governors of the Federal Reserve System.* **ETA Macro: A Model of Energy-Economy Interactions,** *Alan S. Manne, Stanford University.* **The Bauxite Cartel in the New International Economic Order,** *Harold J. Barnett, Washington University.*

Volumes 1 and 2 of Advances in the Economics of Energy and Resources were published under the editorship of Robert S. Pindyck, Sloan School of Management, Massachusetts Institute of Technology

Volume 3, Economic Aspects of New Technology
1980, 274 pp. $63.50
ISBN 0-89232-175-X

CONTENTS: Introduction, *John R. Moroney.* **Worldwide Production Costs for Oil and Gas,** *M.A. Adelman and Geoffrey L. Ward, Massachusetts Institute of Technology.* **Probabilistic Methods for Estimating Undiscovered Petroleum Reserves,** *Frank M. O'Carroll, British Petroleum Corporation, London and James L. Smith, University of Illinois.* **The Nuclear Fuel Cycle and the Demand for Uranium,** *Eugene A. Kroch, Columbia University.* **The Impact of Restrictions on the Expansion of Electric Generating Capacity,** *Edward A. Hudson, Dale W.*

Jorgenson Associates, Dale W. Jorgenson, Harvard University, and David C. O'Connor, Stanford University. **Subsidizing Solar Energy: A Policy Proposal,** *Samuel M. Berman and Anthony C. Fisher, University of California, Berkeley.* **The Assimilation of New Technology: Economic Versus Technological Feasibility,** *S. Charles Maurice and Charles W. Smithson, Texas A&M University.* **Energy, Residuals, and Inefficiency: An Engineering-Econometric Analysis of Environmental Regulations,** *Raymond J. Kopp, Resources for the Future.* **Input Substitution and Biased Technological Change in Resource Intensive Manufacturing Industries,** *Alden L. Toevs, University of Oregon.*

Volume 4, Formal Energy and Resource Models
1982, 275 pp. $63.50
ISBN 0-89232-215-2

CONTENTS: Introduction, *John R. Moroney.* **Energetic Dogma, Energetic Economics, and Viable Technologies,** *Nicholas Georgescu-Roegen, Vanderbilt University.* **The Dynamic Production Potential of an Exhaustible Resource Substitute,** *Nancy T. Gallini, University of Toronto.* **Western Water Rights—Will They Constrain Synthestic Fuels?,** *Constance M. Boris, University of Michigan.* **Combined Energy Models,** *William W. Hogan, John F. Kennedy School of Government and John P. Weyant, Stanford University.* **Economic Growth, Energy Alternatives, and Environmental Protection,** *Eugene Kroch, Columbia University.* **Neoclassical Measurement of Ex Ante Resource Substitution: An Experimental Evaluation,** *Raymond J. Kopp, Resources for the Future, Inc. and V. Kerry Smith, University of North Carolina.* **The Specification and Measurement of Technical Change in U.S. Manufacturing,** *Ernst R. Berndt and David O. Wood, Massachusetts Institute of Technology.* **Empirical Tests of Economic Rent in the U.S. Copper Industry,** *Margaret E. Slade, Federal Trade Commission.*

Volume 5, Econometric Models of the Demand for Energy
1984, 210 pp. $63.50
ISBN 0-89232-327-2

CONTENTS: Introduction, *John R. Moroney.* **Aggregate Consumer Expenditures on Energy,** *Dale W. Jorgenson, Harvard University and Thomas M. Stoker, Massachusetts Institute of Technology.* **Individual Energy Expexnditures,** *Dale W. Jorgenson, Harvard University and Thomas M. Stoker, Massachusetts Institute of Technology.* **Residential Energy Demand in the United States: Introduction and Overview of Alternative Models,** *Lester D. Taylor, University of Arizona, Gail R. Blattenberger, University of Utah and Robert K. Rennhack, Yale University.* **Residential Energy Demand in the United**

States: Empirical Results for Electricity, Lester D. Taylor, University of Arizona, Gail R. Blattenberger, University of Utah and Robert K. Rennhack, Yale University. **Residential Energy Demands in the United States: Empirical Results for Natural Gas,** Lester D. Taylor, University of Arizona, Gail R. Blattenberger, University of Utah and Robert K. Rennhack, Yale University. **Modeling the Aggregate Demand for Electricity: Simplicity Versus Virtuosity,** Ernst R. Berndt, Massachusetts Institute of Technology. **The Residential End-Use Energy Planning System: Simulation Model Strucure and Empirical Analysis,** Andrew A. Goett, Cambridge Systemtics, Inc. and Daniel McFadden, Massachusetts Institute of Technology.

Volume 6, 1987, 220 pp. $63.50
ISBN 0-89232-584-4

CONTENTS: Introduction, John R. Moroney. **Productivity Growth from Total and Variable Cost Functions: An Application to Electric Utilities,** Scott J. Callan, Bentley College. **Capacity Utilization and the Impacts of Pollution Abatement Capital Regulation and Energy Prices,** Catherine Morrison, Tufts University. **Econometric Analysis of Industrial Energy Use in Developing Countries,** Robert Halvorsen, University of Washington. **Interindustry Differences in the Effects of Energy Price-Induced Capital Utilization on Multifactor Productivity Measurement,** Ernst R. Berndt and David O. Wood, Massachusetts Institute of Technology. **Two Stage Budgeting and Consumer Demand for Energy,** Dale W. Jorgenson, Harvard University, Daniel T. Slesnick, University of Texas-Austin, and Thomas M. Stoker, Massachusetts Institute of Technology. **A Dynamic Theory of Resource Development,** Donald A. Hanson, Argonne National Laboratory, Argonne, Illinois and Seung-Dong Lee, University of Alabama-Birmingham. **An Integrated Regional Petroleum Model,** John R. Moroney, Texas A&M University and Dale S. Bremmer, Arkansas State University.

JAI PRESS INC.
55 Old Post Road - No. 2
P.O. Box 1678
Greenwich, Connecticut 06836-1678
Tel: 203-661-7602